THE LATE DEVONIAN MASS EXTINCTION
THE FRASNIAN/FAMENNIAN CRISIS

Critical Moments in Paleobiology and Earth History Series
David J. Bottjer and Richard K. Bambach, Editors

Critical Moments in Paleobiology and Earth
  History Series
David J. Bottjer and Richard K. Bambach, Editors

Mark A. S. McMenamin and Dianna Schulte McMenamin,
  The Emergence of Animals: The Cambrian Breakthrough

Douglas H. Erwin, *The Great Paleozoic Crisis: Life and
  Death in the Permian*

Donald R. Prothero, *The Eocene-Oligocene Transition: Paradise Lost*

Betsey Dexter Dyer and Robert Alan Obar, *Tracing the History of
  Eukaryotic Cells: The Enigmatic Smile*

Perspectives in Paleobiology and Earth History Series,
David J. Bottjer and Richard K. Bambach, Editors

Anthony Hallam, *Phanerozoic Sea-Level Changes*

The Columbia University Press Advisory Committee
for Paleontology:
DAVID J. BOTTJER, CHAIR
RICHARD K. BAMBACH
DAVID L. DILCHER
NILES ELDREDGE
S. DAVID WEBB

The concept of these series was suggested by Mark A. S. and Dianna
L. Schulte McMenamin whose book, *The Emergence of Animals*,
was the first to be published.

The Late Devonian
Mass Extinction
THE FRASNIAN/FAMENNIAN CRISIS

GEORGE R. McGHEE, JR.

COLUMBIA UNIVERSITY PRESS

NEW YORK

Columbia University Press
New York   Chichester, West Sussex
Copyright © 1996 Columbia University Press
All rights reserved
Library of Congress Cataloging-in-Publication Data

Library of Congress Cataloging-in-Publication Data
McGhee, George R.
   The Late Devonian mass extinction : the Frasnian/Famennian crisis
/ George R. McGhee, Jr.
      p.   cm.
   Includes bibliographical references (p.   –   ) and index.
   ISBN 0-231-07504-9 (cloth : acid-free paper). — ISBN
0-231-07505-7 (paper)
   1. Extinction (Biology)   2. Paleontology—Devonian.
3. Catastrophes (Geology)   I. Title.
QE721.2.E97M39   1996   95-38419
575'.7—dc20   CIP

Casebound editions of Columbia University Press books are printed on
permanent and durable acid-free paper.
Printed in the United States of America
c 10 9 8 7 6 5 4 3 2 1
p 10 9 8 7 6 5 4 3 2 1

TO MARAE,
Gràdh mo chridhe thu.

CONTENTS

PREFACE

IT WAS 1977, and I was a graduate student doing research in Germany. Standing in a church tower high above the surface of the Earth, I surveyed the medieval city of Nördlingen below, the table-top flat fields of farmland beyond the city walls, and the line of hills in the distance. I was, in fact, standing in the floor of an enormous impact crater. The line of hills in the distance formed a perfect circle if viewed from the air. They were the remnants of the rim of the Ries Crater, and the flat farmland below was the crater floor. Fifteen million years ago an object fell out of the darkness of space surrounding the Earth, blasting a 24-km-diameter crater in central Europe. It punched straight through 600 m of sedimentary rock cover to vaporize the hard crystalline basement rocks below and ejected 155 km^3 of debris into the atmosphere, some of which would eventually fall more than 400 km away in what is now the Czech Republic (Hörz 1982). After the impact, an enormous plume with coherent downwinds formed over the center of the crater, a cloud "similar to those produced by the rise of a nuclear fireball" (Newsom et al. 1990).

All this was produced by the impact of an object only 1–2 km in diameter (Hörz 1982). And I was wondering if such an impact, on a vastly greater scale than the Ries Crater impactor, had triggered the massive death of animals and plants around the world that we now know as the Late Devonian mass extinction.

My fascination with the Late Devonian mass extinction went back to my 1974 paleontologic fieldwork as a masters student at the University of North Carolina at Chapel Hill. For my thesis research I was attempting to reconstruct the marine communities of organisms that had existed in the ancient Appalachian Sea during the Late Devonian, 367 million years ago. In climbing around the mountains of West Virginia and Maryland, examining the fossil remains of long-extinct animals, I often wondered what had caused the mysterious disappearance of so many of these marine species at the end of what is known as the Frasnian Stage of geologic time. In 1976, as a doctoral student at the University of Rochester (working on a different research project), I still mused over the Late Devonian extinction in discussions with my dissertation adviser, Dave Raup. Dave was at that time bombarding the Earth with asteroids (via a computer, of course) to see how many species would be killed, given varying lethal blast areas. I spent one summer gathering biogeographic data for Dave's project, which most people in the Geology Department considered to be slightly crazy, a research long shot that would probably lead nowhere.

We both knew that seven years before, in 1969, the Canadian paleontologist Digby McLaren had suggested, in his presidential address to the Paleontological Society, that the Late Devonian mass extinction might have been triggered by the impact of an asteroid. His suggestion was dismissed by most scientists as being an intellectual throwback to the old days of geologic catastrophism. Others assumed he was simply joking. Dave Raup took the idea seriously. Some four years later, following the publication of the famous Alvarez study in 1980—with its implication that the dinosaurs had perished at the end of the Cretaceous because of the impact of an asteroid with the Earth —many others began to take the "catastrophic" idea seriously, as well.

What caused the Late Devonian mass extinction? Was it indeed triggered by the impact of an extraterrestrial object? It has been two decades now since I first pondered those questions. These years have taken me many places around the world, and I have spent many long hours with colleagues who, like myself, are fascinated with the mystery of the Late Devonian biotic crisis. In this book I have set down some of their ideas, and many of my own, concerning the puzzle of this mass extinction. I would like to acknowledge here the helpful research assistance (in the field, in the lab, in logistics, or in words of encouragement) of R. K. Bambach, U. Bayer, P. Bultynck, J. M. Dennison, E. G. Kauffman, D. J. McLaren, W. A. Oliver, Jr., E. J. Olsen, C. J. Orth, D. M. Raup, P. Sartenaer, A. Seilacher, J. J. Sepkoski, Jr., O. H. Walliser, and W. Ziegler. I also gratefully acknowledge the many conversations, debates, and sometimes arguments with my colleagues (listed in roughly chronological order) C. W. Stock, R. G. Sutton, Z. P. Bowen, D. L. Woodrow, G. C. Baird, P. W. Signor, R. E. Plotnick, P. Copper, M. R. House, J. Kullmann, J. Wendt, D. E. G. Briggs, D. Jablonski, J. E. Sorauf, J. T. Dutro, Jr., C. W. Stearn, S. E. Scheckler, M. Streel, G. Klapper, N. D. Newell, A. J. Boucot, S. M. Stanley, A. Hallam, Hou Hongfei, W. T. Holser, G. Racki, J. Kalvoda, R. Feist, J. Hladil, N. M. Farsan, R. T. Becker, R. Cuffey, A. Raymond, C. E. Brett, M. L. McKinney, E. Schindler, C. A. Sandberg, J. W. Valentine, W. T. Kirchgasser, F. Langenstrassen, K. Wang, H. H. J. Geldsetzer, M. M. Joachimski, W. Buggisch, and M. Narkiewicz. Not all these colleagues will agree with what I have written here; in fact, I can imagine a few who will vehemently disagree.

Finally, I thank my wife, Marae, who throughout these years has amusedly tolerated my fascination with a catastrophe that happened so very long ago.

THE LATE DEVONIAN MASS EXTINCTION
THE FRASNIAN/FAMENNIAN CRISIS

Critical Moments in Paleobiology and Earth History Series
David J. Bottjer and Richard K. Bambach, Editors

Crises in the History of Life

THE FIVE PHANEROZOIC CRISES

The ultimate measure of crises in the history of life is the loss of species diversity from one geologic time interval to another. Extinction is a normal part of the evolution of life and is inevitable. All species eventually become extinct. Life continues, however, as species normally give rise to descendant lineages. Thus the diversity of species, while fluctuating, is maintained through the passage of geologic time. Life goes on.

At certain periods in the history of the Earth the diversity of life has decreased catastrophically. These changes are referred to as "mass extinctions," to distinguish them from the normal extinctions that occur constantly but that do not markedly affect the diversity of life, as the lost species are replaced by new species.

The simplest, and oldest, measure of mass extinction is a plot of the diversity of life as a function of geologic time. (See table 1.1 for the geologic timescale.) In 1860 John Phillips, then professor of geol-

Table 1.1. The Geologic Timescale, Giving the Eras, Periods, and Epochs of the Phanerozoic Eon (the past 570 million years) [*]

Era	Period	Epoch	Time (Ma BP)
Cenozoic	Quaternary	Holocene	
		Pleistocene	
	Neogene	Pliocene	
		Miocene	
	Paleogene	Oligocene	
		Eocene	
		Paleocene	65
Mesozoic	Cretaceous	Late (Gulf)	
		Early	
	Jurassic	Late (Malm)	
		Middle (Dogger)	
		Early (Lias)	
	Triassic	Late	
		Middle	
		Early (Scythian)	245
Paleozoic	Permian	Late (Zechstein)	
		Early (Rotliegendes)	
	Carboniferous	Late	
		Early	
	Devonian	Late	
		Middle	
		Early	
	Silurian	Late (Ludlow/Pridoli)	
		Middle (Wenlock)	
		Early (Llandovery)	
	Ordovician	Late (Bala)	
		Middle (Dyfed)	
		Early (Canadian)	
	Cambrian	Late (Merioneth)	
		Middle (St. David's)	
		Early (Caerfai)	570

[*] The timescale is constantly being refined; where proposed, new epoch names are given in parenthesis. Ages at the base of each era from Harland et al. (1989).

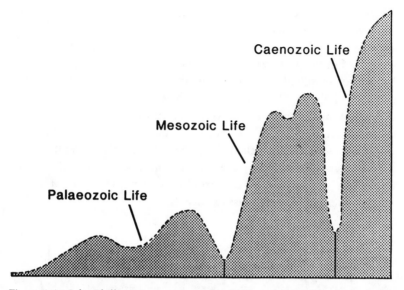

Figure 1.1. John Phillips's eras, proposed in 1860, were based on the recognition of two major periods of biotic crisis in the Earth's history. Note that he also recognized two smaller, less catastrophic, diversity declines within the middle of the Mesozoic and the middle of the Paleozoic (the Late Devonian). (Modified from Rudwick 1976.)

ogy at Oxford, published a rough estimate of the diversity of life (figure 1.1) showing a relative increase in diversity with time. More important, he illustrated periods of decline in diversity as well, particularly the major losses in diversity that marked the close of his proposed Paleozoic and Mesozoic Eras. Phillips's eras were in essence defined on the basis of mass extinction. Our earliest measurement of the history of life is based on death.

Much more refined data are now available to us than those used by Phillips more than 130 years ago. Much of what we know today about the magnitude and effect of mass extinctions in geologic time is due to the work of one paleobiologist, Jack Sepkoski. Sepkoski (1982a,b) has compiled a massive amount of data on the diversity of life at the familial level for the past six hundred million years, pulling together the work of generations of paleontologists who have gone before. Their labor has born fruit. Plotting familial diversity as a function of time reveals five major crisis periods in Earth history, when diversity decreased markedly (figure 1.2). These crises occur in the terminal Ordovician, the Late Devonian, the terminal Permian, the terminal Triassic, and the terminal Cretaceous. These severe diversity losses in the past are referred to as mass extinctions (a concept to be further explored in more detail in chapter 2) and are referred to colloquially as "The Big Five" among paleontologists. The mass extinction most popularly known is that of the end-Cretaceous, as it is then that the last of the nonavian dinosaurs perish. The end-Cretaceous event, also called the Cretaceous/Tertiary event, is not, however, either the first or even second most severe of these events (table 1.2). The single most catastrophic event in the history of life on Earth is the end-Permian mass extinction.

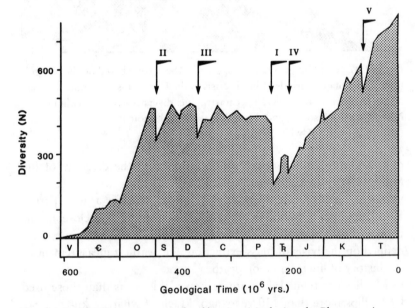

Figure 1.2. The five major periods of biotic crisis during the Phanerozoic. The curve plots the diversity of animal families in the Earth's oceans as a function of geologic time. Major diversity losses, marked by arrows, occurred in the terminal Permian (I), the terminal Ordovician (II), the Late Devonian (III), the terminal Triassic (IV), and the terminal Cretaceous (V). Following each crisis, a geologically significant span of time elapsed before marine life recovered to previous levels of diversity (delineated by the bars from the arrows above). Modified from Erwin et al. (1987).

Table 1.2. Magnitudes of the Major Crises in the
History of Life on Earth. Data from Sepkoski (1982a). *

Mass extinction	Marine families eliminated (%)
I. end-Permian	50
II. end-Ordovician	22
III. Late Devonian	21
IV. end-Triassic	20
V. end-Cretaceous	15

* For each event the percentage of families of marine animals which
perish is given.

In the end-Permian crisis more than 50% of all the animal families
present in the Earth's oceans perish. As many as 83% of the marine
genera are eliminated (Sepkoski 1986), and it is estimated that 96%
of all the species in the oceans vanish (Raup 1979). Perhaps only 4%
of the species present in the late Permian world survive into the
Triassic. The remaining mass extinctions are much less severe, based
on the magnitude of familial diversity loss (table 1.2). In all cases,
however, a geologically significant amount of time elapses before the
numbers of new species originations are able to bring diversity back
to previous levels (figure 1.2).

THE UNUSUAL LATE DEVONIAN CRISIS

The Late Devonian event is particularly controversial at present. Indeed, one evolutionary biologist has bemoaned, "The Late Devonian
extinction is one where not even the major facts are agreed on yet"
(Van Valen 1984). Note that four of the "Big Five" mass extinctions
listed in table 1.2 are indicated to occur at the end of a geologic
period, but that one, the Devonian Period, is listed as "Late" instead.
This is due to the fact that the most severe diversity loss in this
extinction event occurs not at the end of the Devonian but *within*
the Devonian.

The Devonian is divided into three epochs (table 1.1), and
these three epochs are further divided into seven stages (table 1.3).
The Late Devonian Epoch occurs thus in two stages, the older

Table 1.3. The Devonian Period Divisions, Giving the Epochs and Stages Within Each Epoch.[*]

Period	Epoch	Stage	Age at base (Ma BP)	Duration
Carboniferous	Early	Tournaisian	362.5	
Devonian	Late			
		Famennian	367.0	4.5 Ma
	Late	Frasnian	377.4	10.4 Ma
	Middle			
		Givetian	380.8	3.4 Ma
	Middle	Eifelian	386.0	5.2 Ma
	Early			
		Emsian	390.4	4.4 Ma
		Pragian	396.3	5.9 Ma
Devonian	Early	Lochkovian	408.5	12.2 Ma
Silurian	Late	Pridoli		

[*] Ages at the base of the stages, and their durations, from Harland et al. (1989).

Frasnian and the younger Famennian. (A brief aside: The word *Frasnian* is pronounced as if it were spelled *Franian;* that is, the s is silent. The pronunciation is French, as one of the early boundary strata for this stage is the base of the *Assise de Frasnes* in southern Belgium.)

The Late Devonian biotic crisis is alternatively referred to as the "Frasnian/Famennian" event, in that the major phase of diversity loss appears to take place not in the terminal stage of the Devonian (the Famennian; see table 1.3), but further back in time, in the stage preceding the Famennian (the Frasnian). Even this point has been disputed, however. Widespread disagreement exists on whether it was an "event" or a series of multiple events, and an even wider diversity of opinion exists regarding the cause of the Frasnian/Famennian crisis. All these disagreements and disputations will be explored in subsequent chapters.

The Late Devonian is a fascinating interval of time. The entire Devonian itself is a very important interval in the evolution of life on the Earth, a noisy and busy time, full of momentous events. The Devonian is often referred to as the Age of Fishes by paleoichthyologists and as the Age of Land Plants by paleobotanists. The first land plants, small and inconspicuous, evolved before the Devonian, gaining a small beachhead in the invasion of land back in the Silurian, but it is in the Devonian that the full onslaught and seizure of the land by plants occurs. The first forests to grace the surface of the Earth evolved in the Middle Devonian, with towering trees reaching into the sky. Those trees were unlike any alive today, however, and the Middle Devonian forests would appear weird and alien to us if we could travel back in time to take a walk in their sylvan quietness. Animal life also made the difficult transition to land in the Devonian. Marine life crept out of the sea and into the freshwater streams and lakes in greater and greater numbers during the Devonian. The arthropods, the dominant form of animal life on the planet, successfully left the protective waters of the Earth to walk in the harsh dry air of the land.

A great event occurs in the Late Devonian, at least from our point of view as land-dwelling vertebrates. Our ancestors also emerged from the streams to face the challenge of life in dry air, to gasp the thin mixture of gases in the atmosphere, and to walk (clumsily, but all first steps are) in the Devonian forests. The first tetrapods, part of our direct lineage, evolved in the Frasnian. And they received a sad welcome, an early horror. Our primitive tetrapod ancestors witnessed the Frasnian/Famennian mass extinction.

THE TIME IN QUESTION

The distinguished biostratigrapher and ammonoid specialist Michael House is upset about time. More specifically, he is upset about what passes for precise time measurement at the geologic stage level, particularly in the ancient Paleozoic: "We have had in recent years a flood of published time scales, often with supposedly precise figures given of radiometric dates for stage boundaries, but the diverse estimates from publication to publication show that these are spurious" (House 1991:392). He then focuses on a particularly irksome point with laser intensity:

Datings are made either on the assumption that stages within
the system are of equal duration or . . . on the assumption that
the number of faunal zones within a stage gives an estimate of
its duration. No biostratigrapher would for one moment support either of these suppositions. Figures resulting from such
calculations are almost useless, and it is surprising that geophysicists should present such figures as if they are meaningful and
that learned publishing houses should print such idle speculation. [House 1991:392]

What could provoke the intensity of such a response? One answer is
the timescales of Harland and colleagues (1982) and (1989), and
Michael House is by no means alone—quite the opposite, in fact,
particularly within the biostratigraphic and paleobiological communities (see, for example, the discussion in the excellent paper of Oliver
and Pedder 1994:179–180). The accurate measurement of geologic
time is of extreme importance in our understanding of the evolutionary processes that have shaped the course of life on Earth. But the
precise measurement of fine-scale units of geologic time is absolutely
crucial to the study of mass extinction.

The numbers given in table 1.3 are impressive at first glance. The
age of the beginning of each stage and the durations of the stages are
given to the nearest one hundred thousand years, or, in the language
of timescale-speak, to an accuracy of 100 ka. Thus one can say that
the first Frasnian dawn came as the sun rose in 377.4 Ma BP (i.e.,
377.4 million years before the present), or that the Famennian lasted
4.5 Ma (or 4.5 million years). Or can one?

Estimates of the Relative Duration of the Devonian Stages

In 1975 the great Devonian biostratigrapher Art Boucot reported the
results of an interesting exercise in time measurement. He wanted to
know what other experts thought about the passage of time within
the Devonian, specifically, how long the Devonian stages are in duration. Biostratigraphers think in relative time, as evolution occurs in a
sequence of species originations and extinctions. Species A first appears here in the stratigraphic column, followed by species B (which

appears higher in the section), followed by species C (still higher), and so on. Assembling the sequence of evolutionary events that occur in the history of life on Earth is the basis of the geologic time scale. The original geologic time scale is a relative time scale, a sequence of events.

One can say with precision in a sequence that event A occurs first, B occurs second, and so on. But that does not tell you when A occurred, or how long a period of time lapsed before B occurred. The absolute measurement of time in geology is based upon radiometric dating, carefully measuring the ratios of radioactive isotopes within a geologic sample and then calculating how long a period of time has lapsed from the formation of the sample based upon the known period of time required for those particular radioactive isotopes to decay. (For further information of radiometric dating, consult any good introductory geochemistry textbook, such as Faure 1986.)

Boucot (1975) asked five biostratigraphic specialists to make a relative assessment of the length of the stages of the Devonian, based upon the evolutionary sequence of events that occur within each of their specialty groups. Biostratigraphers are reluctant to attempt such an exercise, as it requires an assumption to be made concerning the relative numbering of events. Does the occurrence of five evolutionary events in stage A and ten events in stage B mean that stage B is twice as long as stage A? To say so would commit one to the assumption that the evolutionary events took place at the same rate, and they may not have. Evolution may have occurred at a faster pace in stage B, and the ten events may have occurred over the same period of time as the five events in stage A.

Art Boucot can be persuasive, however, and five specialists agreed to take part in the exercise. Those specialists (and their groups of study) were M. R. House (ammonoids), G. Klapper (conodonts), J. T. Dutro, Jr. (brachiopods), W. A. Oliver, Jr. (corals), and A. R. Ormiston (trilobites). Their estimates were reported in a list of "raw numbers" (Boucot 1975:68–69), which I have converted to percentages and which are given in table 1.4.

Of most importance here is the relative duration of the two stages of the Late Devonian. The five biostratigraphers estimate the Frasnian Stage to span between 13.3% and 18.7% of the total duration of the Devonian Period (table 1.4A). Their estimates of the duration of the

Table 1.4. Estimates of the Relative Duration of Devonian Stages[*]

A. Biostratigraphic estimates of stage durations, made by 5 specialists, as reported by Boucot (1975)

	M.R.H.	G.K.	J.T.D.	W.A.O.	A.R.O.
Famennian	23.0%	22.9%	20.4%	15.0%	14.6%
Frasnian	15.3	18.7	14.2	13.3	16.6
Givetian	15.3	12.5	14.2	16.6	18.7
Eifelian	15.3	12.5	10.2	16.6	12.5
Emsian	7.7	12.5	20.4	8.3	20.8
Pragian	15.3	8.3	10.2	16.6	4.2
Lochkovian	7.7	12.5	10.2	13.3	12.5

B. Timescale of Harland et al.

	(1982):	(1989):	Average of specialist's estimates
Famennian	14.6%	9.8%	19.2%
Frasnian	14.6	22.6	15.6
Givetian	12.5	7.4	15.5
Eifelian	14.6	11.3	13.4
Emsian	14.6	9.5	13.9
Pragian	14.6	12.8	10.9
Lochkovian	14.6	26.5	11.2

[*] See text for discussion; specialists listed are M. R. House (M.R.H.), G. Klapper (G.K.), J. T. Dutro, Jr. (J.T.D.), W. A. Oliver, Jr. (W.A.O.) and A. R. Ormiston (A.R.O.).

Famennian is much more variable, from 14.6% to 23.0%. Four of the specialists are in agreement that the Famennian is longer than the Frasnian, with estimates that the Famennian is 1.7% to 7.7% longer than the Frasnian. Only one specialist has the Frasnian 2.0% longer than the Famennian.

In 1982 a radiometric timescale, based upon the examination of thousands of data, was published by a committee of scientists from Cambridge University in England (Harland et al. 1982). As a radiometric timescale, it was ultimately to be based upon measurements of radioactive isotope ratios taken from geologic samples scattered throughout the stratigraphic record, and was expected to yield absolute (not relative) time determinations. Devonian workers awaited its publication and opened its pages with anticipation, only to express dismay or outrage. In the final analysis, the simple assumption had been made that all the stages of the Devonian were of equal duration, and a linear extrapolation had been made from two radiometric "tie points" (see Harland et al. 1982), one in the Early Carboniferous and one in the Silurian, with the Devonian in between. All the stages are reported as being 7 Ma in duration (6 Ma for the Givetian, but that is due to rounding). The result is that all the stages comprise equally 14.6% of the time span of the Devonian (12.5% for the Givetian, again due to rounding), as shown in table 1.4B. I remember one colleague muttering at the time that the timescale of Harland and colleagues (1982) was not worth the paper upon which it was printed.

Clearly, more work was needed. A new radiometric timescale was issued in 1989 by the Cambridge working group (Harland et al. 1989). In the intervening seven years, five new "tie points" had been determined for the Devonian (see Harland et al. 1989), and the stages were now determined to be of unequal durations (the timescale given previously in table 1.3). The resultant relative durations of the Devonian stages, using the new timescale, are given in table 1.4B, and the average of the biostratigraphic estimates is given in a column adjacent for comparison of the two.

The Harland and colleagues (1989) stages are now of very unequal duration indeed, ranging from a low of only 7.4% of the total Devonian (the Givetian) to a high of 26.5% (the Lochkovian). The biostratigraphic average estimate has an inequality range only from 10.9% to 19.2% (table 1.4B).

Of most importance here is the great disparity in the estimated durations of the Late Devonian stages. One stage is fully 12.8% longer than the other, but it is the Frasnian that is determined to be the longest and the Famennian the shortest! Only one biostratigraphic estimate has the Famennian shorter than the Frasnian, and that by only 2.0%.

The general opinion among Devonian paleontologists is that there is something very wrong here. House (1991:392–393) writes of the Harland and colleagues (1989) timescale:

They gave a bizarre length for the Lochkovian ... well over twice any informed estimate. Their Frasnian figure (22.6%) is again well above any other estimate, yet the Frasnian has been reduced in length (both at the top and base) in recent years by decisions of the Commission on Stratigraphy.

Estimates of the Absolute Duration of the Devonian Stages

In 1985 a symposium was held at Princeton University in honor of Professors A. G. Fischer and F. B. Van Houten, both of whom had pioneered research into cyclic phenomena in global sedimentary processes. Much attention was paid to the Milankovitch cycles, which are produced by variations in the orbit of the Earth around the sun, and their effect upon the Earth's climate and eventually the stratigraphic record. The implications of the detection of such extraterrestrially induced cycles on the Earth were clear to all at the symposium. Such orbital cycles were regular, periodic, clocklike. If they could be detected in the stratigraphic record, we could produce a geologic timescale of a precision never before dreamed of.

My colleague Ulf Bayer, on a research visit from the University of Tübingen, was with me at Rutgers University at the time, and we drove down to Princeton to give a short presentation of some of our research into temporal cycles and the stratigraphic record. In our project we had determined to turn the traditional research procedure on its head, to do the exact opposite of what was normally expected. Rather than carefully analyze the stratigraphic record with the hope of detecting the existence of a periodic signal amid all the noise in the data, we started with the a priori assumption that a periodic signal existed in the data and we would *adjust the stratigraphic record itself* to best fit the signal. We carefully measured changes in sedimentary thickness, facies types, and biological diversity in the Devonian stratigraphic record of New York State. We came up with a series of units, or groupings, of physical and biological data that were of unequal duration, of differing time spans, according to current timescale measurement. We then made the assumption that *our units were originally of equal duration* and adjusted the Devonian timescale accordingly, compressing it a bit here, expanding it a bit there (Bayer and McGhee 1986, 1989).

The results of our study are given in table 1.5, to be contrasted with the Harland and colleagues (1982; 1989) timescales. The two biggest temporal discrepancies in the comparisons are the durations of the Famennian and the Givetian in our timescale versus that of Harland and colleagues (1989). Our estimate of the Famennian is 8.5

Table 1.5. Estimates of the Absolute Duration of the Stages of the
Devonian, Contrasting the Physically and Biologically Derived Estimates

Stage:	Using physical criteria		Using biological criteria	
	Harland et al. (1982)	Harland et al. (1989)	Bayer and McGhee (1986, fig 12)	Ziegler and Sandberg (1990)
Famennian	7 Ma	4.5 Ma	13 Ma	10 Ma
Frasnian	7	10.4	9	5
Givetian	6	3.4	12	
Eifelian	7	5.2	7	
Emsian	7	4.4	6	
Pragian	7	5.9	4	
Lochkovian	7	12.2	9	

Ma longer and our estimate for the Givetian is 8.6 Ma longer. House
(1991) also considers the Harland and colleagues (1989) estimate for
the Givetian to be unreasonably short.

The Famennian duration story is a bit more complex, however. In
our original cycle analysis, the Famennian came out as 13 Ma in
duration (Bayer and McGhee 1986, figure 12). Ulf was uneasy with
this determination, as it would push the base of the Carboniferous to
a considerably younger position. We argued the point, but for the
final publication Ulf prevailed and rather than "following the cyclic
analysis without control in the Carboniferous" (see discussion in
Bayer and McGhee 1986:400) the reported duration of the Famennian is given as 6 Ma rather than 13 Ma (Bayer and McGhee 1986,
table 2). That shorter estimate is still 1.5 Ma longer than the later
Harland and colleagues (1989) figure. In table 1.5 I have chosen "to
boldly go with the cyclic analysis without control in the Carboniferous" and give the original analysis determination of a 13 Ma duration
for the Famennian.

Is there any other supporting evidence for such a determination?
Yes, and from a perhaps surprising source. House (1991:392) states
that "no biostratigrapher would for one moment support the assumption" of equal biological zone durations. Yet Ziegler and Sandberg
(1990), both biostratigraphers of international fame, have done just
that for the Late Devonian. Willi Ziegler, Charlie Sandberg, and

Table 1.6. The Late Devonian Standard Conodont Zonation*
From Ziegler and Sandberg (1990).

Stage	Zone	Time
TOURNAISIAN (Carboniferous)		+ 10 Ma
FAMENNIAN	Late *praesulcata* Middle *praesulcata* Early *praesulcata*	+ 9
	Late *expansa* Middle *expansa*	+ 8
	Early *expansa* Late *postera*	+ 7
	Early *postera* Late *trachytera*	+ 6
	Early *trachytera* Latest *marginifera*	+ 5
	Late *marginifera* Early *marginifera*	+ 4
	Late *rhomboidea* Early *rhomboidea* Latest *crepida*	+ 3
	Late *crepida* Middle *crepida*	+ 2
	Early *crepida* Late *triangularis*	+ 1
FAMENNIAN	Middle *triangularis* Early *triangularis*	0
FRASNIAN	*linguiformis* Late *rhenana*	− 1
	Early *rhenana* *jamieae*	− 2
	Late *hassi* Early *hassi*	− 3
	punctata *transitans*	− 4
	Late *falsiovalis*	

Stage	Zone	Time
FRASNIAN	Early *falsiovalis* (pars)	– 5
GIVETIAN	Early *falsiovalis* (pars)	

* The measurement of time above and below the Frasnian/Famennian boundary, in zonal increments of one million years, is modified from the timescale of Ziegler and Sandberg (1990).

their colleagues are responsible for the construction of the conodont zonation of the Late Devonian which is now known as the Standard Conodont Zonation (see table 1.6). It will be used throughout this book, and other biostratigraphic zonations will be referenced with respect to the Late Devonian Standard. For the point at hand, however, note that twenty-two zones exist in the Famennian and only ten exist in the Frasnian. If (and it is indeed an if) the zones are of roughly equal duration, then the amount of time represented by the Famennian is about twice that of the Frasnian.

Ziegler and Sandberg (1990) have gone further, however, and have proposed that each conodont zone is about a half million years (0.5 Ma) in duration, with some exceptions. Some zones they consider, on the basis of their long experience in the pattern of Late Devonian conodont evolution, to have been shorter than 0.5 Ma (the *praesulcata* zonal interval, for example). They proposed a timescale for the Late Devonian of 15 Ma, of which the Famennian spans 10 Ma and the Frasnian 5 Ma (table 1.5). This compares favorably with the proportion of time allotted to each stage from the original cycle analysis of Bayer and McGhee (1986), though our absolute duration estimates are 3 Ma longer for the Famennian and 4 Ma longer for the Frasnian.

More important, note that Harland and colleagues (1989) also consider the Late Devonian to have been around fifteen million years in duration (14.9 Ma, actually). Their estimation of the duration of the stages within the Late Devonian is almost the exact opposite of that of Ziegler and Sandberg (1990), however. We thus have two diametrically opposed proposals for the relative length of the Famennian and Frasnian Stages, one the virtual reverse of the other. Which one is correct? It is impossible to say at present, though a hint does exist that the Late Devonian timescale of Harland and colleagues (1989) may eventually be proven to be wrong. Using a totally different type of calibration analysis, Fordham (1992) has made a determination of 13 Ma for the duration of the Famennian, the exact same number that Ulf Bayer and I produced in our cycle analysis back in 1985.

Last, return to the Late Devonian Standard Conodont Zonation given in table 1.6. Against the original zonal positions I have marked off the one-million-year increments proposed by the Ziegler and Sandberg (1990) timescale. I have numbered these a little differently from the original timescale, which numbered the years in negative time back from the datum of the Famennian/Tournaisian (Devonian/ Carboniferous) boundary. The subject of this book is the Frasnian/ Famennian mass extinction. Thus I have taken the Frasnian/Famennian boundary as the datum in table 1.6. Time then is measured in negative terms before that horizon (back in time), and in positive terms after that horizon (forward in time).

SUMMARY: TIMESCALES MATTER

Timescales have serious theoretical implications. They are not simply games with numbers. As I mentioned at the beginning of this section, the accurate measurement of geologic time is of extreme importance to our understanding of the evolutionary process that has shaped all life on the planet. Ziegler and Sandberg (1990) have proposed a timescale, a set of numbers, but these numbers have evolutionary implications. Their timescale implies that the evolution of Late Devonian conodont species was regular and clocklike. The biological process of genetic change and mutation within these living species was like the clocklike process of radioactive decay of nonliving isotope species. Is this tenable? Yes, there are indeed neutralist models of evolution that would fit such a scenario.

What of the Harland and colleagues (1989) timescale? What are its theoretical implications? This timescale would require a rapid increase in the rate of conodont species evolution in the Famennian. A zone would have an average duration of 1.04 Ma in the Frasnian and would decrease by almost an entire order of magnitude to an average duration of only 0.20 Ma in the Famennian. That is, the time it took to produce an evolutionarily distinct set of zonal species in the Frasnian was 1,040,000 years, whereas that same amount of evolutionary change occurred in only 200,000 years in the Famennian. Is this tenable? Yes, the fossil record abounds with examples of such rapid bursts of evolution, of dramatic increases in evolutionary rate of speciation.

Thus we cannot eliminate either timescale on the basis of their theoretical requirements. I would like to stress the words *theoretical requirements* here. We are not dealing with idle speculation. Whether the Ziegler and Sandberg (1990) or the Harland and colleagues (1989) timescale is ultimately proved correct, or indeed some other timescale to be proposed in the future, certain *theoretical consequences must follow from that timescale* concerning the evolutionary process of conodont speciation, and of all other life, in the Late Devonian.

Last, I mentioned at the beginning of this section that the precise measurement of fine-scale geologic time units is absolutely crucial in the study of mass extinction. As seen in this section also, however, we still do not have consensus on such a timescale. Thus we shall have to do the best that can be done with what is present to work with. I have chosen to devise a "crisis interval timescale" around the critical Frasnian/Famennian boundary, which is given in table 1.7. The absolute datum for the Frasnian/Famennian boundary I have taken from the Harland and colleagues (1989) timescale; it stands at 367.0 Ma BP. Time measurement before and after this datum is based on the timescale of Ziegler and Sandberg (1990), and the resultant ages determined for each zone are given in table 1.7. I have also given the correlation of the Standard Zonation to that of the earlier zonation of Ziegler (1971), which was in use in the 1980s and which will be encountered later when we explore the events that unfolded in the search for asteroid impact evidence at the Frasnian/Famennian boundary (chapter 9).

This "crisis interval timescale" spans some 3.0 Ma around the Frasnian/Famennian boundary (table 1.7). It will be used repeatedly for this critical interval of time throughout the book. The Standard Conodont Zonation (table 1.6) is of importance for correlation within the Late Devonian around the world and for correctly placing

Table 1.7. Timescale for the Crisis Interval Surrounding the Frasnian/ Famennian Boundary*

Time (years BP)	Crisis Interval Conodont Zonations:	
	Late Devonian Standard (Ziegler and Sandberg 1990)	Ziegler (1971)
ca 365.5 Ma	Early *crepida*	Lower *crepida*
ca 366.0 Ma	Late *triangularis*	Upper *triangularis*
ca 366.5 Ma	Middle *triangularis*	Middle *triangularis*
ca 367.0 Ma FAMENNIAN	Early *triangularis*	Lower *triangularis*
FRASNIAN ca 367.5 Ma	*linguiformis*	Uppermost *gigas*
		Upper *gigas*
ca 368.0 Ma	Late *rhenana*	
		Lower *gigas*
ca 368.5 Ma	Early *rhenana*	

*This timescale will be used in subsequent chapters. Age of the Frasnian/Famennian boundary from the timescale of Harland and colleagues (1989), ages above and below this datum based on the duration estimate of 0.5 Ma per conodont zone in the timescale of Ziegler and Sandberg (1990). The earlier zonation of Ziegler (1971) is given as a reference, as it was in use in the early stages of the worldwide search for bolide impact evidence (see chapter 9).

events that happened at zonal horizons other than the crisis interval surrounding the Frasnian/Famennian boundary. The Harland and colleagues (1989) timescale (table 1.3) will be used when discussing events that occur on a broader time frame, earlier in the Devonian or later in the Carboniferous. Thus we shall see tables 1.3, 1.6, and 1.7 again in later chapters.

II

Severity of the
Mass Extinction

TIMES OF "NORMAL" EXTINCTION

How does one measure an extinction event? The answer to that
question is, on the one hand, not simple, and, on the other hand,
simple. Diversity loss is the single fundamental signal of ecological
crisis, the true measure of extinction. At one point in time many
exotic, colorful, noisy, active life forms exist; at a later point there is
the quietness of a world with just a few species left. Species go extinct
all the time, however, and that is a normal part of the evolution of
life. Thus we need somehow to distinguish "abnormal" extinction
from "normal" extinction, to differentiate a healthy ecosystem from
one in crisis.

The Metrics of Mass Extinction

In principle, we can simply compare the diversity of life in one geologic time interval with the following interval. At the boundary between the two time intervals, three things can happen. A number of species present in the first interval may go extinct, which we can count as N_1. The remainder of species (those that do not go extinct) survive into the second interval, which we can count as N_2. Last, new species may originate in the second interval that were not present in the first, which we can count as N_3. The simplest measure of extinction is thus

$$E = N_1 \qquad (2.1)$$

The metric E is the *absolute extinction magnitude*. Easy? Not quite. Suppose we have a loss of one million species in one extinction event and a loss of one thousand species in another. The former event seems more severe than the latter, but is it? Perhaps not. Suppose further that in the first event ten million species were present before the extinction and one million were lost. The proportion, or percentage, of species lost in the first event is thus 10%. In the second event, suppose that only two thousand species were present before the extinction and that one thousand perished in the event. The proportion of species lost in the second event is thus 50%. Clearly, the second event can now be seen to be more severe than the first. A world that has lost a tenth of its living diversity is bad, but a world that has lost fully one-half is far worse. Absolute magnitudes may thus be misleading, and we can correct for this by scaling the absolute extinction magnitude against the total number of species present at the time of the extinction event:

$$E/N = N_1/(N_1 + N_2) \qquad (2.2)$$

The metric E/N is the *proportional extinction magnitude,* and it is easily convertible to percentages by scaling against 100, if desired.

But this is not quite all. Suppose we encounter two periods in geologic time during which ten thousand species are lost. As you gathered earlier, these losses may not be of equal severity. Suppose the loss of the first ten thousand species occurred over ten million years; thus about one thousand species were lost each million years or so. And suppose the loss of the second ten thousand species occurred in only one million years, or less. A world in which environmental stresses degrade ecosystems relentlessly for millions of years, never abating as species after species slowly succumbs, is bad. A world that experiences an environmental catastrophe of a magnitude large enough to exterminate a vast number of species virtually instantaneously is worse. Clearly, the second event is more severe than the first; thus the rate at which species are lost is as important as the number of species that are lost. To take this into account, we can scale the absolute extinction magnitude against a time interval Δt, usually taken to be a million years:

$$E/\Delta t = N_1/\Delta t \tag{2.3}$$

The metric $E/\Delta t$ is known as the *absolute extinction rate*. As we saw earlier, absolute magnitudes may be misleading; thus we can apply the same scaling correction to the absolute extinction rate (equation 2.3) as we applied to the absolute extinction magnitude (equation 2.2), or

$$E/(N\Delta t) = N_1/(N_1 + N_2)\Delta t \tag{2.4}$$

This final extinction metric, $E/(N\Delta t)$, is known as the *proportional extinction rate*.

There is more to ecosystem dynamics than just extinction, however, just as there is more to population dynamics than just death. In population dynamics, as much attention is paid to birth rate as to death rate. All animals eventually die; that is inevitable. A healthy population, however, has death rates that are offset, or surpassed, by birth rates, so that the population maintains its numbers of individuals or actually increases in size (to a limit). In geologic time, we track the number of species originations (O) as well as the number of extinctions (E), where

$$O = N_3 \tag{2.5}$$

This metric is known as the *absolute origination magnitude*.

For each of the extinction metrics, a similar metric for origination is obtained by simply substituting O for E and N_3 for N_1 in equations 2.2 to 2.4. The result is metrics of absolute origination magnitude (equation 2.5), proportional origination magnitude, absolute origination rate, and proportional origination rate.

To return to the point stressed at the beginning of the chapter, the ultimate measure of ecological crisis is diversity loss. Global changes in diversity are a function of the rate at which new species are added to the Earth's biosphere and the rate at which old species are lost from the Earth's biosphere; that is,

$$\Delta N/\Delta t = O/\Delta t - E/\Delta t \qquad (2.6)$$

The metric $\Delta N/\Delta t$ is known as the *absolute turnover rate,* the rate at which diversity is changing in an ecosystem over geologic time. An alternative measure of ecosystem dynamics is the *proportional turnover rate,* or $\Delta N/N\Delta t$, which is calculated using proportional origination and extinction rates, rather than absolute ones. Scaling against standing diversity is sometimes used to double-check conclusions based solely on absolute magnitudes (see the discussion for equation 2.2).

Healthy ecosystems are in equilibrium, with turnover rates at zero, or with small positive and negative fluctuations about zero, in geologic time. Sharp negative values for turnover rates indicate the rapid loss of species diversity, a sign in geologic time that something has seriously gone wrong. Note carefully that such negative values can be produced in two quite different ways in equation 2.6. The most obvious way is to elevate extinction rates sharply. A less obvious way is to decrease origination rates sharply. The same result is obtained, however. In the analysis of the evolution of ecosystems, *it is a mistake to consider the pattern of extinction rates alone.* And it may be misleading in the analysis of mass extinction as well, as will be discussed in chapter 3.

TIMES OF "ABNORMAL" EXTINCTION

How severe must an extinction be in order to be a "mass" extinction? The simplest answer would be the loss of at least 15% of the planet's standing diversity of life at the familial taxonomic level in a geologic time period of no more than 15 Ma, as this is in fact how the "Big Five" are distinguished (Sepkoski 1982a; see chapter 1).

This simple definition has two key requirements: The loss of diversity must be very large (-15% at the familial level), and the diversity loss must occur in a geologically short period of time (no more than 15 Ma). Fifteen percent may not seem at first like a very large number, but it is if one considers the shape of the hierarchical system used in the classification of life. It is like a pyramid, small at the top but larger and larger toward the base. The base unit in the classification of life is the species, a common gene pool of interbreeding (i.e., genetically compatible) individuals. Usually, several species are then grouped together to comprise a single genus, and several genera are grouped together to comprise a single family. Thus to kill one family, several genera must be eliminated, which requires the termination of numerous species, which requires the death of hundreds of thousands to millions of individuals. And that is just for one taxonomic family.

Similarly, fifteen million years (maximum) may seem like a long period of time. On geologic timescales, however, it is quite small. The entire Phanerozoic, the Age of the Kingdoms Animalia and Plantae, spans some 600 Ma or so (table 1.1). A 15 Ma duration represents 2.5% of the entire Phanerozoic time span, a mere 2.5% of time in which the accumulated multimillion-year evolution of 15% of all families vanishes. And that is the maximum outside range—the mass extinction may (and usually does) occur in a much shorter period of time. The consequences of variable time durations (Δt) and mass extinction magnitudes will be explored in more detail in chapter 3.

Jablonski (1986a) considers mass extinction to consist of three important components: magnitude, duration, and breadth. To the qualities of magnitude and duration discussed earlier, he has added "breadth." What is breadth? It is more of a qualitative concept— mass extinction must be ecologically and geographically broad. A mass extinction strikes across broad ecological categories, affecting the sea as well as the land, decimating herbivores as well as carnivores, plants as well as animals. A mass extinction is not confined to one geographic area or region, a continent here or an ocean there. The entire planet is affected.

Raup (1991a) characterizes mass extinction in terms of rare events. That is, normal extinction occurs all the time; thus it is not rare. Mass extinction is rare in the history of life. But what is normal extinction? To answer this question, Raup analyzed the time ranges of a sample of 17,621 fossil genera to determine their pattern of survivorship. The results of those analyses were then interpolated to the species level to give the species kill curve of figure 2.1.

Raup found that 39% of all extinctions that occur in the geologic record have a proportional species kill magnitude of 5% or less.

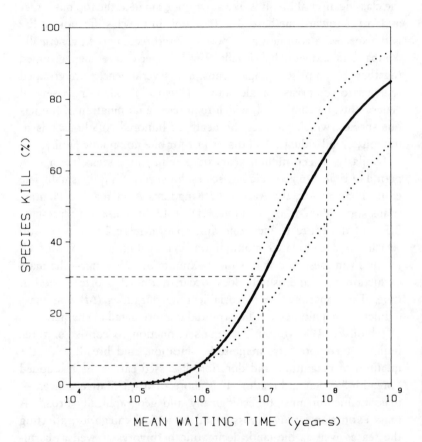

Figure 2.1. The species kill curve of Raup (1991a). The proportional extinction magnitude of species killed is plotted here versus the mean waiting time. For a given percentage of species killed in an event, read over to the solid line and down to the expected waiting time for an event of that magnitude. Species kills of 5% to 10% have mean waiting times of 1.0 Ma or so; species kills of more than 60% (mass extinction) are rare, with expected waiting times of 100 Ma. Dotted curves indicate the uncertainty range about the position of the solid curve. From Raup (1992a), used with permission of the author.

Extinction events in which more than 60% of the species perish account for only 4% of the extinctions that occur in the approximately 600 Ma span of the Phanerozoic. Between these extremes Raup has constructed the kill curve of figure 2.1 by asking the additional question, "How frequently should an event of a given magnitude occur in time?" given the observed time and frequency distribution of small and large extinctions in the fossil data. From this analysis we can see that extinction events in which 5% to 10% of the species vanish are quite common in geologic time and generally occur (have a mean waiting time of) every million years or so. They are thus very common in time and are a normal part of the evolution of life.

Extinction events with species kill magnitudes of more than 60% are quite rare—only 4% of the observed extinctions in the Phanerozoic. Raup calculates that they have a mean waiting time of 100 Ma or more; that is, they are rare in time as well. They are uncommon; they are abnormal; they are mass extinction.

The Frasnian/Famennian Diversity Crisis

Estimates of the severity of the Frasnian/Famennian crisis are given in table 2.1. These estimates are given for three separate taxonomic levels in the classification of life: the familial, the generic, and the species. From the preceding discussion of the numbers-pyramid effect of the hierarchical system in the classification of life, one would expect the numbers to become larger and larger at lower and lower taxonomic levels, and that expectation is confirmed in table 2.1.

The numbers in table 2.1 are given in the proportional extinction magnitude metric (see the previous discussion of metrics). More variability is present in the estimates of the magnitude of familial extinctions and less occurs in the estimates of generic- and species-level diversity losses. The variability of the familial estimates is due partially to history and partially to differing calculations in computation of the statistics.

Newell (1967) compiled familial data in the middle 1960s largely because the only globally correlatable data then available existed only at the familial level. Those data also had a time resolution only at the level of the geologic epoch. Thus Newell's estimate includes all

Table 2.1. Estimates of the Severity of the Late Devonian Mass Extinction (in the proportional magnitude metric)*

Taxonomic Level	Diversity Loss Estimate (%)	Reference
Families	− 30	Newell (1967)
	− 13	Valentine (1969)
	− 14	Raup and Sepkoski (1982)
	− 21	Sepkoski (1982a)
	− 38	McKinney (1985)
Genera	− 60	Boucot (1975)
	− 50	Sepkoski (1986)
	− 55	Jablonski (1991)
Species	− 70	McGhee (1982)
	− 82	Jablonksi (1991)

* Note the effect of the taxonomic hierarchy; for a given number of families to go extinct, an even larger number of genera, and an even larger number of species, must perish.

extinctions that occur in the Late Devonian Epoch and does not resolve whether those extinctions occur in the beginning, middle, or end of the Epoch. Thus his estimate of the severity of the Frasnian/ Famennian mass extinction is too high. On the other hand, a similar compilation of up-to-date data on familial extinctions, also resolved only at the epoch level, yielded an even higher statistic of diversity loss, that of − 38% (McKinney 1985). The difference between Newell's statistic of − 30% in 1967 and McKinney's statistic of − 38% in 1985 is due to the vast improvement in our knowledge concerning the distribution of families in geologic time, largely due to the work of Sepkoski (1982b). McKinney's statistic, like Newell's, contains extinctions that occur throughout the Late Devonian, not just those at the end of the Frasnian, and is too high an estimate.

Lower estimates may result from imprecision in calculation of the metric. For example, the − 14% statistic of Raup and Sepkoski (1982) was computed solely by measuring the drop in standing diversity from the Frasnian Stage to the Famennian. The problem with this calculation is that new species originations in the Famennian stage are included in the total standing diversity of that time interval, thus making the drop in diversity at the Frasnian/Famennian boundary seem smaller than it actually was. To be specific: The denominator used in the calculation of the proportional extinction magnitude (equation 2.2) should be $(N_1 + N_2)$. In the calculation of the -14% statistic, the denominator used was $(N_1 + N_2 + N_3)$, and thus the estimate is too low. The estimate of Sepkoski (1982a) was computed using equation 2.2, and it is our best measure of the severity of familial diversity loss that occurred in the Frasnian/Famennian mass extinction.

Estimates of the magnitude of generic diversity loss are fairly tight and are in agreement that somewhat over half of the genera present on the Earth did not survive the Frasnian/Famennian mass extinction (table 2.1). The two estimates of the magnitude of species diversity loss are also quite close, which is interesting, as they were obtained in totally different fashions. The statistic of Jablonski (1991) is a model estimate, obtained by extrapolation from generic data. It is thus subject to possible error from two sources: the accuracy of the generic data used in the calculation and the accuracy of the model parameters used in the extrapolation. The estimate of McGhee (1982) is empirical; it is based on a count of the number of species of fossilized marine animals that disappear from the rock record in the Frasnian/Famennian extinction in eastern North America. A global data count, at the species level, does not exist at present. Thus a count of the number of species that go extinct in another area of the world will no doubt yield a different estimate. In addition, as an empirical measure it is probably an underestimate. Empirical measurements from the stratigraphic record are almost always underestimates, because of the imperfect nature of fossil preservation. Thus the -70% figure is a minimum estimate; we know that *at least that proportion of species vanished* from the fossil record, but the actual number was probably higher.

The Frasnian/Famennian crisis clearly meets the magnitude requirements of a mass extinction designation. It is "mass" indeed. Probably three-quarters of all species then alive on the Earth did not survive the mass extinction. That elimination of three-quarters of the diversity of life cascades up the taxonomic hierarchy, and the echo of that devastation at the species level is seen in the grim statistics: More than half of all the genera of the planet lost and more than a fifth of all the families.

III

Duration of
the Mass Extinction

DATA RESOLUTION AND DURATION ESTIMATES

We saw in the previous chapter that the duration of a diversity loss crisis in geologic time is just as important as magnitude in the analysis of mass extinction. A mass extinction is unusually big, and it is unusually fast. In chapter 2 we examined what is known of the magnitude of the Frasnian/Famennian mass extinction (E and E/N); in this chapter we will explore the time duration of the event (Δt).

The temporal duration of the Frasnian/Famennian mass extinction has been, and still is, the subject of considerable debate. The duration debate has also been linked often with the debate concerning the causation of the extinction event, with those favoring gradual mechanisms of extinction arguing for a protracted event and those favoring catastrophic mechanisms arguing for an abrupt event. The two questions are, however, at least semi-independent. For example, if a gradual process of global climatic deterioration were rapidly accelerated by an asteroidal impact, the result in the stratigraphic record would still be an event of significant temporal duration. To use this duration result to rule out a catastrophic component of causation would, of course, be mistaken, as the causation (as hypothesized) was multifactorial and not simple.

Estimates of the duration of the mass extinction have varied widely. Some published duration estimates are 15 Ma (Raup and Sepkoski 1982), 10 Ma (House 1967), 6–8 Ma (Farsan 1986), 7 Ma (McGhee 1982), 3–5 Ma (Copper 1984), 1.3 Ma (McGhee 1988a), 1.0 Ma (two conodont zones; Ziegler 1984), 0.5 Ma (one conodont zone; Sorauf and Pedder 1984), less than 0.5 Ma (McLaren 1982, 1984), to less than 0.1 Ma (Sandberg et al. 1988). This is an astonishing range of published opinion: from a duration estimate of fifteen million years at one extreme to less than one hundred thousand years at the other. And that is just the beginning.

Among those workers who maintain that the event is not of short (i.e., instantaneous) duration, disagreement exists as to whether the event records a series of episodic extinctions ("stepdown extinctions," Copper 1986; "episodic gradualism," Kalvoda 1986), a series of multiple extinctions with distinctive and recurrent phases (House 1985; Becker 1986), the simple accumulation of successive extinctions (Farsan 1986), or the gradual disappearance of species (Dutro 1984). It is no wonder, faced with such a great variety of opinion in the published scientific literature, that one prominent evolutionary paleobiologist has thrown up his hands and exclaimed, "The Late Devonian extinction is one where not even the major facts are agreed on yet" (Van Valen 1984:130).

Duration estimates are strongly influenced by the completeness of the stratigraphic record and the minimum interval of stratigraphic correlation. Luckily, abundant stratigraphic sequences exist around the globe for the Late Devonian interval. Rather than stratigraphic completeness, the main problem in determining the duration of the Frasnian/Famennian event has been in the correlation resolution of the data.

EPOCH-LEVEL RESOLUTION

An early work that pointed out that a major extinction event took place during the Devonian is that of Newell (1967), as discussed in chapter 2. His analyses indicate that approximately 30% of all families of marine organisms became extinct in the Late Devonian (figure 3.1), a peak that rises considerably above other extinction levels in the earlier Devonian and later Carboniferous.

The data available to Newell (1967) could only be obtained at the epoch level of temporal resolution and at the familial level of taxonomic resolution. Plotting these data as points at epoch termini (figure 3.1), the maximum extinction appears to occur at the very end of the Devonian. Thus the extinction appears to mark the close of the Devonian and the boundary to the Carboniferous. If that were truly the case, then the title of this book would have been *The End-Devonian Mass Extinction* and not the unusual *Late Devonian* mass extinction (see chapter 1). We now know that is not the case, and that the main pulse of the extinction occurs around the Frasnian/Famennian boundary *within* the Late Devonian, and not at the Devonian/Carboniferous boundary. Interesting events do take place at the latter boundary, however, as we shall see a little further on in this chapter.

It was not possible to show this, on an international level, in 1967 with temporal resolution at the epoch level. Strictly speaking, the -30% pulse in familial diversity loss could occur at the very beginning of the Late Devonian, in the middle, or at the very end (or anywhere in between, for that matter). The data would simply allow one to say that a 30% loss in families occurs somewhere in the Late Devonian, and the point placed at the end of the epoch in figure 3.1 could just as well have been placed at the beginning or the middle.

Newell (1967) chose to present his analyses using a magnitude metric, with no measure of time duration involved (the calibration of the Paleozoic timescale in the 1960s was much worse than even the present uncertain state). The Late Devonian has a duration of 15 Ma (or 14.9 Ma, in the Harland et al. 1989 timescale). Thus the minimum limit of resolution, or error of estimate, of the point in question in figure 3.1 is 15 Ma. Imagine the point repositioned in the middle of

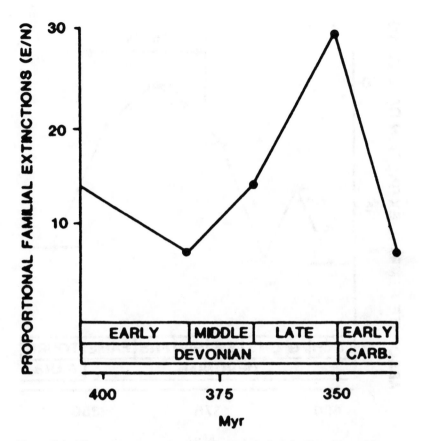

Figure 3.1. The apparent pattern of extinction during the Devonian with time resolution at the epoch level. The proportional extinction magnitude for marine families of animals is here plotted at the ends of epochs, using the timescale of Raup and Sepkoski (1982). (Data from Newell 1967. Modified from McGhee 1988a.)

the Late Devonian, with an error bar of ± 7.5 Ma on either side. That was the state of knowledge in 1967.

FINER: STAGE-LEVEL RESOLUTION

In the intervening years data of finer temporal resolution, though still at the familial taxonomic level, have been gathered by Sepkoski

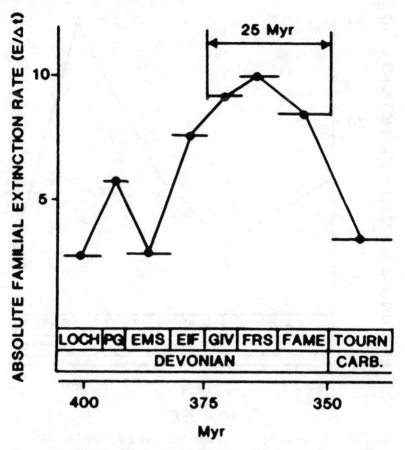

Figure 3.2. The apparent pattern of extinction with time resolution at the stage level. The absolute extinction rate of marine families is plotted versus the timescale of Raup and Sepkoski (1982). Devonian stage names are abbreviated; see table 1.3 for a full list. (Data from Sepkoski 1982b, modified from McGhee 1988a.)

(1982b) and analyzed by Raup and Sepkoski (1982). These data are largely resolvable to the stage level, though a few remain that are only known to the epoch level (Sepkoski 1982b). In presenting their analysis results, Raup and Sepkoski chose to use a rate metric rather than a magnitude metric and to go with absolute numbers rather than with proportional ones (figure 3.2).

In their analyses the Frasnian Stage emerges with the maximum extinction rate for the entire Devonian, not the Devonian/Carboniferous boundary (figure 3.2). Note, however, that the points for the temporally flanking Givetian and Famennian stages are also high and are close to the value of the Frasnian peak (figure 3.2). In their analyses, Raup and Sepkoski (1982) concluded that all three stages— the Givetian, Frasnian, and Famennian—were above the computed "background" extinction rate, with the Frasnian at the maximum. One seemingly could conclude from figure 3.2 (and from Raup and Sepkoski 1982) that extinction rates are elevated for the entire span of time from the Givetian though the Famennian, or a period of some 25 Ma. Thus, though the temporal duration of the extinction event in the Newell (1967) data remained unresolvable—the event could have been of very short duration somewhere within the Late Devonian, or of extended duration spread throughout the Late Devonian—the stage-level data of Sepkoski (1982b) appear to indicate that there is no short-duration extinction event, but rather an extended period of high extinction rates in the latter half of the Devonian Period. Extinction rates rise in the Givetian Stage, peak in the Frasnian, begin to fall in the Famennian (but are still high), and subside in the earliest Carboniferous (figure 3.2). The analysis of data only resolvable to the stage level has also led some workers to conclude that the Late Devonian mass extinction is not "statistically significant" from background (Raup and Sepkoski 1982; Hubbard and Gilinsky 1992).

Thus the end result of the Raup and Sepkoski (1982) analysis is curious. The data have a much finer time resolution, to the stages within epochs, and not just to the epoch level. The Late Devonian extinction event seems to expand in time, however, just the reverse of what one would expect from finer time resolution. In the Newell (1967) data, the extinction had a maximum estimated duration of the Late Devonian Epoch, or 15 Ma. In the Raup and Sepkoski (1982) analysis, the extinction encompasses both stages within the Late Devonian Epoch and the Givetian Stage of the Middle Devonian Epoch, for a combined estimated time duration of 25 Ma.

Most important, however, is the emergence of the Frasnian Stage as the maximum, the peak, in the Raup and Sepkoski (1982) analysis. Whatever extinction-producing event was going on during the Middle and Late Devonian was at its worst in the Frasnian.

STILL FINER: SUBSTAGE-LEVEL RESOLUTION

Finer-scale temporal data are scarce at the present, particularly on a global geographic scale. The data that do exist are generally confined to taxonomic groups that are important for international geologic correlation, such as ammonoids and conodonts in the case of the Devonian. Multispecies data, resolvable to substage levels, usually exist only for restricted geographic areas at present. On the other hand, finer-scale temporal data are generally obtainable at the species taxonomic level. That has been a general rule in the nature of paleontologic data that is only now breaking down. Data internationally correlatable on a global scale have had only coarse temporal and high taxonomic resolution. Data obtainable at the local level have very fine relative temporal resolution and species-level taxonomic resolution. The problem has been, and still is, correlating events seen in fine scale at each individual local level to each other across great geographic distances. But it is being done, slowly but surely, because of the ongoing efforts of many successive groups of scientists in the International Geological Correlation Program.

The substage-level data that do exist indicate that the single broad extinction peak seen in figure 3.2 is an artifact of the level of temporal resolution. When examined in still finer time detail, the single continuous peak breaks down into three separated peaks. There is a separate peak in extinction rates *within* the Givetian, *within* the Frasnian, and *within* the Famennian (figure 3.3). Each peak is flanked on either side by times of lower extinction rate values. These three separated peaks, resolvable at the substage level, are "blurred" into a continuous peak if only stage-level resolution is applied to the data. Try it for yourself: Simply draw a line that connects the high points, or peaks, on the curves given in figure 3.3 from the Givetian to the Frasnian to the Famennian. The result will be an extended continuous period of high extinction rates, from the Givetian through the Famennian, for the conodonts. A smaller, Givetian through Frasnian, broad peak results for the brachiopods and ammonoids. They are artifactual, do not exist, and are the result of a coarser level of time resolution.

Note that the late Frasnian now stands out as an isolated period of elevated extinction rates, flanked by lower values in the earlier Fras-

EXTINCTION RATES

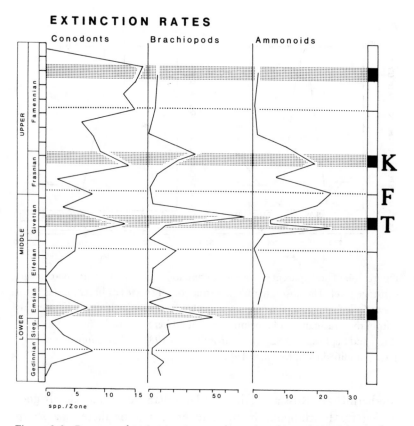

Figure 3.3. Patterns of extinction in conodonts, brachiopods, and ammonoids during the Devonian with time resolution at the substage level. Metric is the absolute rate of extinction, calculated in number of species per zone (rather than per Ma). In addition, letters in the far right margin indicate the temporal position of three of the ammonoid extinction events of House (1985): T = Taghanic event, F = Frasnes event, K = Kellwasser event. (Modified from Bayer and McGhee 1986.)

nian and in the early Famennian, in the figure 3.3 data. The Frasnian peak in figure 3.2 now has become a *late* Frasnian peak in figure 3.3. Is it real, or is it artifactual?

Let us examine the late Frasnian peak in more detail with a different set of data. Figure 3.4 gives data on brachiopod extinctions from two separate regions of the Earth, New York State (USA) and the

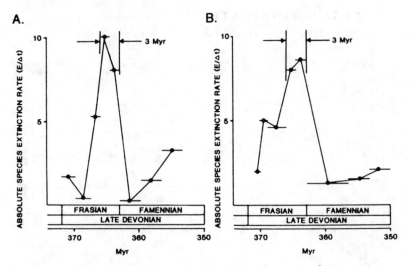

Figure 3.4. The apparent pattern of extinction with time resolution at the substage level. The absolute rate of extinction of species of brachiopods in New York State (A) and in the southern Urals of Russia (B) for time intervals within the Frasnian and Famennian is plotted here versus the timescale of Raup and Sepkoski (1982). (Data from Dutro 1981 and Stepanova et al. 1985, modified from McGhee 1988b.)

southern Urals (Russia), in the Late Devonian. Something clearly goes wrong for brachiopods in the late Frasnian, for there is a sharp elevation of the rate of species extinctions in both regions of the world. The extinction rate of brachiopod species is lower in the temporally flanking early Frasnian and early Famennian, as was also seen in the data of figure 3.3.

Note that in these data the late Frasnian peak consists of two points, not just one (figure 3.4). This means there are two distinct substage stratigraphic units in which large numbers of brachiopod species vanish, in both regions of the world. The highest extinction rate of brachiopod species actually occurs before the end of the Frasnian in New York State, whereas the highest rate in the southern Urals is in the very end substage unit of the Frasnian. This variability in timing of the maximum pulse of extinction may be real, in that the brachiopod faunas of the two regions of the world may have reacted differently to whatever environmental stress was triggering the species

extinctions. That is, the New York State brachiopods may have suc-
cumbed in greater numbers early in the ecological crisis, whereas the
southern Ural brachiopods may have resisted extinction for a longer
span of time, only to succumb finally in the very end of the Frasnian.
On the other hand, the difference in timing of the maximum peak in
the two regions of the world could be entirely artifactual, simply due
to incompleteness in the fossil record.

The important point is the existence of two substage stratigraphic
units with high extinction rates in the late Frasnian in two widely
separated regions of the world. That is, the stratigraphic "two-unit"
signal is real and stands out from any preservational "noise" or
incompleteness in total number of species found as fossils in the
two regions.

The existence of two points in the curves of figure 3.4 indicates (1)
that the crisis that struck these brachiopod species in the Late Devonian had a maximum duration of 3.0 Ma and (2) that the crisis came
in the late Frasnian.

EVEN FINER YET: BIOZONAL-LEVEL RESOLUTION

Quantitative data are very rare at present for the finest level of
international time resolution, that of the zone. Data that do exist are
usually for conodonts or ammonoids, zonally important groups. In
the analysis of House (1985), five separate diversity crises are seen to
strike the ammonoids during the Late Devonian. He has further
narrowed the time of occurrence of these diversity crises down to the
zone. The temporal position of each of the five ammonoid diversity
crises, or events, is given with reference to the Standard Conodont
Zonation in table 3.1.

Three of the five events are of greater magnitude than the remaining two (House 1985): the Frasnes event, Kellwasser event, and
Hangenberg event. Note that these three occur at stage boundaries:
the Givetian/Frasnian boundary, Frasnian/Famennian boundary, and
Famennian/Tournaisian (Devonian/Carboniferous) boundary. (See Simakov 1993 and Wang et al. 1993 for a discussion of events at the
last horizon.) Thus we again see in House's data the three-point
structure that underlies what appears to be a continuous 25 Ma peak

Table 3.1. Zonal Positions of Five Ammonoid Extinction Events that Occur in the Late Devonian*

Stage	Zone	Ammonoid Extinction Event (House 1985)
TOURNAISIAN (Carboniferous)		
FAMENNIAN	Late *praesulcata* Middle *praesulcata* Early *praesulcata*	■ Hangenberg event
	Late *expansa* Middle *expansa*	
	Early *expansa* Late *postera*	
	Early *postera* Late *trachytera*	■ Annulata event
	Early *trachytera* Latest *marginifera*	
	Late *marginifera* Early *marginifera*	■ Enkeberg event
	Late *rhomboidea* Early *rhomboidea* Latest *crepida*	
	Late *crepida* Middle *crepida*	
	Early *crepida* Late *triangularis*	
FAMENNIAN	Middle *triangularis* Early *triangularis*	
FRASNIAN	*linguiformis* Late *rhenana*	■ Kellwasser event
	Early *rhenana* *jamieae*	
	Late *hassi* Early *hassi*	

Stage	Zone	Ammonoid Extinction Event (House 1985)
	punctata	
	transitans	
FRASNIAN	Late *falsiovalis* Early *falsiovalis* (pars)	■ Frasnes event
GIVETIAN	Early *falsiovalis* (pars)	

* Data from House (1985, 1989). Note that two discrete events occur *within* the Frasnian, and three *within* the Famennian.

of extinction at the stage level of time resolution (figure 3.2). These events can be narrowed down to separated zones, however, lasting an estimated 0.5 Ma.

The late Frasnian ammonoid diversity crisis is listed as the Kellwasser event in table 3.1. The name for the event is taken from the Kellwasser Limestone horizons, a stratigraphic interval in which black shales and bituminous limestones were deposited across a widespread area of Europe and northern Africa. Ammonoids suffer a major loss of species during the deposition of these peculiar anoxic sediments. We will encounter the important Kellwasser limestones many times throughout this book.

The Kellwasser event is listed as occurring in the *linguiformis* Zone (table 3.1), which is the latest Frasnian. Was it a single event, confined to one zone? House (1985) mentions the possibility that a separate, smaller loss of ammonoid diversity (the *Crickites* event) may occur later in the Early *triangularis* Zone; that is, in the earliest Famennian.

Let us now focus on the Frasnian/Famennian boundary, using the crisis interval timescale developed in chapter 1. Data indicate that biologically significant events occur throughout the latest three zones of the Frasnian, and the earliest three zones of the Famennian (table 3.2), or a span of time of some 3.0 Ma. The severest pulse of extinction does appear to occur in the *linguiformis* Zone, which contains not only the ammonoid diversity crisis of the Kellwasser event, but also the conodont diversity crisis. Only three species of the important

Table 3.2. The Apparent Pattern of Extinction with Time Resolution at the Zonal Level.*

Time (years BP):	Biological Event:
Early *crepida* Zone ca 365.5 Ma	■ Final extinction of corals and stromatoporoids in Moravia (Hladil et al. 1986)
Late *triangularis* Zone ca 366.0 Ma	
Middle *triangularis* Zone ca 366.5 Ma	■ Decimation of reefs in the Urals (Kalvoda 1986) ■ Final and total extinction of all atrypoid brachiopods (Copper 1986)
Early *triangularis* Zone ca 367.0 Ma FAMENNIAN	■ Decimation of calcareous foraminifera in eastern Europe (Kalvoda 1986) ■ Final and total extinction of all cricoconarids (Schindler 1990a)
FRASNIAN	
linguiformis Zone ca 367.5 Ma	* SEVEREST PULSE OF EXTINCTION MAGNITUDES ■ Conodont diversity crisis horizon (Sandberg et al. 1988) ■ Global decimation of rugose corals (Sorauf and Pedder 1986) ■ "Kellwasser event" global decimation of ammonoids (House 1985, 1989)
Late *rhenana* Zone ca 368.0 Ma	■ Beginning of cricoconarid extinctions (Schindler 1990a) ■ Beginning of coral and stromatoporoid extinctions in Moravia (Hladil et al. 1986) ■ Decimation of reefs throughout western Europe (Tsien 1980) ■ Beginning of atrypoid brachiopod extinctions (Copper 1986)
Early *rhenana* Zone ca 368.5 Ma	■ Beginning of extinction of reefs in Europe (Sandberg et al. 1988)

* Age of the Frasnian/Famennian boundary from Harland and colleagues (1989); ages of the zones before and after this datum are based on the estimate that each zone is 0.5 Ma in duration (Ziegler and Sandberg 1990).

zonal genera *Palmatolepis* and *Polygnathus* survive this crisis (Sandberg et al. 1988). Significant extinctions occur in the earlier Late *rhenana* Zone, however, not the least of which is the decimation of reefal ecosystems throughout Europe (table 3.2).

In summary, data at the zonal level of time resolution indicate that multiple events occur around the Frasnian/Famennian crisis interval, and not just one single, isolated, geologically instantaneous event. Narrowing down the focus further, a continuous series of events appears to occur in sequence from the Early *rhenana* Zone to the Middle *triangularis* Zone, a break occurs in the Late *triangularis* Zone, with further extinctions then taking place later in the Early *crepida* Zone.

The main sequence of events, from Early *rhenana* to Middle *triangularis* zones, occurs over a maximum duration of only 2.0 Ma (table 3.2). The two most severe pulses appear to occur in the *linguiformis* Zone (the maximum diversity loss horizon) and in the preceding Late *rhenana* Zone. The maximum duration of this two-point peak is only 1.0 Ma, using the timescale of table 3.2.

The late Frasnian two-point peak of high extinction rates in brachiopod species was estimated to be of a maximum duration of 3.0 Ma, using data with substage-level time resolution (figure 3.4). Data with zonal levels of time resolution allow us to narrow that figure down to 2.0 Ma. If the Late *rhenana* and *linguiformis* zones are taken as the two points of the late Frasnian extinction peak, the Δt can be narrowed still further, to 1.0 Ma.

FINEST POSSIBLE?—SUBZONAL-LEVEL RESOLUTION

The most severe pulse of extinction in the Frasnian/Famennian crisis occurs in the *linguiformis* Zone. Does this pulse represent one event or several? To answer that question requires a time resolution finer than the zone itself. No such resolution exists at present; we have reached the limit of international time correlation.

We have not reached the minimum limit of time resolution at the local level, however. Eberhard Schindler has conducted an extensive analysis of the biological events that occur in the Upper Kellwasser Limestone, that is, the horizon of the later part of the *linguiformis*

Zone (Schindler 1990a,b; 1993). He has examined this critical horizon, the very latest Frasnian, in numerous stratigraphic sections in
Germany, France, and Morocco.

Schindler has found that five separate biological events occur in the
Upper Kellwasser horizon, which are correlatable over distances of
110 km in Europe (figure 3.5). The first to occur is the disappearance
of characteristic Frasnian species of trilobites (horizon A in figure
3.5). The second event is not an extinction at all, but a noticeable
increase in population densities of entomozoan ostracodes and homoctenid cricoconarids (horizon B). This increase in densities of these
small invertebrates is brought to an abrupt halt at a higher horizon
(C in figure 3.5), in which the entomozoans disappear and the homoctenids suffer sharp reductions in population densities of individuals.

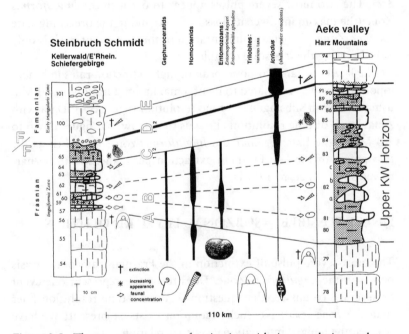

Figure 3.5. The apparent pattern of extinction with time resolution at the
subzonal level. Shown is the correlation of a series of five biological events
(A–E) in the latest Frasnian and earliest Famennian, which occur in the same
sequence in localities some 110 km apart in Germany. (From Schindler 1990,
1993. Art courtesy of E. Schindler.)

At this same horizon characteristic Frasnian graphoceratid ammonites disappear, but peculiar large bivalve molluscs briefly appear as fossils. Horizon D is the Frasnian/Famennian boundary itself, at which an increase in population densities of *Icriodus* conodonts (shallow water species) occurs. The homoctenid cricoconarids, which manage barely to survive horizon C in the Frasnian, finally disappear totally at horizon E in the earliest Famennian.

What is the conclusion of this high-resolution, microstratigraphic study? It is this: The main pulse of the Frasnian/Famennian mass extinction appears not to be a bedding plane event, but a five-bedding plane event. However, these five bedding planes occur within a unit that is only 0.5 to 1.0 m thick (figure 3.5). It is difficult to calculate with any degree of certainty how much time this 0.5–1.0-m-thick unit represents, certainly much less than 0.5 Ma, as the unit occupies only the later part of the *linguiformis* Zone and does not span the entire zone. Sandberg and colleagues (1988) estimate the duration of deposition of the Upper Kellwasser Limestone at the Steinbruch Schmidt site to be 0.1 Ma, or one hundred thousand years, at the most. The most severe bedding plane event, in terms of disappearance of species or sharp reductions in population densities of individuals within species, occurs at horizon C (figure 3.5). Sandberg and colleagues (1988) consider the 5 cm bed at horizon C to represent no more than 21,500 years at the very most.

Do the same five events occur in other parts of the planet? That is a very crucial question. It could be argued that the five events seen in the Steinbruch Schmidt section are artifacts of preservation. That is, a single bedding plane event occurs (say horizon C), and horizons A and B preceding it are artifacts of first, not finding trilobites above horizon A in this section even though they are still alive, and second, better preservational conditions at horizon B that produce an apparent increase in numbers of individuals in populations that actually does not occur. On the other hand, the fact that the same five events, in the same sequence, occur in the same Upper Kellwasser Limestone 110 km away at the Aeketal section provides strong evidence against the preservational artifact argument.

The key, therefore, is correlation, as it always is. If the same five events can be correlated around the planet, then clearly they are real. We cannot say if they are globally correlatable at the moment, though work is in progress at various sites around the world. Peculiar geo-chemical events occur at the Upper Kellwasser horizon as well, in addition to the biological events outlined earlier. These physical events will be outlined in detail in chapter 9, where we will encounter the now famous Steinbruch Schmidt site again.

EVOLUTIONARY DYNAMICS OF THE CRISIS

In chapter 2 it was argued that it is a mistake to consider extinction rates alone in the analysis of the evolution of ecosystems, particularly the global degradation and collapse of ecosystems that occurs in mass extinction. The real measure of biological crisis is diversity loss, and diversity loss may also be driven by the cessation of new species originations—in essence, the failure to replace species lost to extinction.

To illustrate this point, let us consider one brief example. There exists a persistent observation among field geologists, going back over one hundred years (see Chadwick 1935 for a summary of New York State field studies), that a marked drop in species diversity occurs at the end of the Frasnian. The fact that extinction rates for New York State brachiopod species are elevated long before the end of the Frasnian (figure 3.4) suggests that the drop in species diversity at the end of the Frasnian is not a simple function of extinction rate magnitudes alone (see McGhee 1982:494). In fact, Frasnian marine ecosystems in New York State appear to have been flourishing (in terms of standing species diversity) during the same interval of time corresponding to the maximum extinction rate value given in figure 3.4 (McGhee and Sutton 1981, 1983, 1985; Sutton and McGhee 1985).

Let us now consider a different metric discussed before, the absolute turnover rate ($\Delta N/\Delta t$), for both the New York State brachiopods and their relatives in the southern Urals. Remember, positive turnover rates indicate an increase in the number of species with time; zero rates, an equilibrium diversity; and negative rates, a loss in species diversity. Sharp negative values indicate an unusually fast loss of species diversity (i.e., a mass extinction).

Sharp negative turnover rates occur for brachiopod species in both

New York State and the southern Urals in the very latest Frasnian (figure 3.6). This abrupt net loss of species diversity is not, however, a direct function of elevated extinction rates (cf. figures 3.4 and 3.6). Extinction rates are high in the last two substage units in the Frasnian in both regions (figure 3.4). The sharp negative pulse in species turnover rate occurs only in the last substage unit of the Frasnian (figure 3.6). The rapid loss of species diversity is a much shorter event, occurring in less than 1.5 Ma (substage level of time resolution).

Curiously, New York State brachiopod species flourish (positive turnover rates, figure 3.6) during the same time interval characterized by elevated extinction rates (i.e., the first point in the late Frasnian peak in figure 3.4). They do so because origination rates of new species are higher, per time interval, than corresponding extinction rates (McGhee 1988a,b; 1990). It is the abrupt reversal in the magnitude of origination versus extinction rates that precipitates the rapid

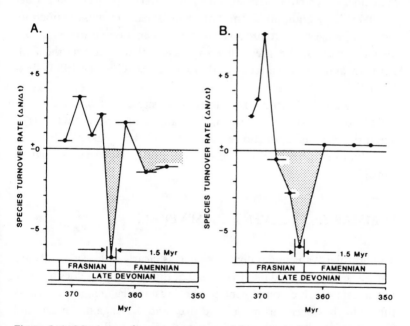

Figure 3.6. Maximum duration estimates of the period of brachiopod species diversity loss in New York State (A) and the southern Urals of Russia (B). Compare with figure 3.4. In both regions an abrupt loss in diversity occurs at the very end of the Frasnian. (Modified from McGhee 1989b.)

loss of brachiopod species diversity seen at the very end of the Frasnian.

Sadly, too few paleobiologists are examining total ecosystem behavior during times of catastrophic diversity loss, in that nearly every analysis in the literature concentrates exclusively on extinction rates. The role of origination rate decline in driving diversity loss, much less the dynamic interplay of the two evolutionary rates, is usually ignored. But times are changing. In an excellent paper, Oliver and Pedder (1994) conduct a thorough evolutionary analysis of diversity crises in the rugose corals during the span of the Devonian. They find that the sharp net losses in rugosan generic diversity that occur at the end of the Frasnian are preceded by earlier extinction highs and that the negative turnover rates (high rates of diversity loss) are driven by a later decline in originations. In comparison to the pattern of diversity dynamics seen in the Devonian brachiopods (discussed earlier), they conclude, "Our coral data fit the same pattern" (Oliver and Pedder 1994:188). In addition to the Frasnian/Famennian mass extinction, evidence suggests that precipitous declines in origination rates are also a factor in the great diversity losses that occur in the end-Permian and end-Cretaceous mass extinctions (Hüssner 1983; Bakker 1977).

In viewing the Frasnian/Famennian diversity crisis from an ecological perspective, one must ask more than the question, "What triggered the elevated extinction rates?" One must also ask, "What was the inhibiting factor that caused the cessation of new species originations?"

SUMMARY: ONE EVENT OR SEVERAL?

The Frasnian/Famennian diversity crisis is clearly not a single instantaneous event. I would like to stress that that conclusion does not rule out a catastrophic component to the Frasnian/Famennian crisis, as this point has been misunderstood over and over again. An asteroid impact may well have triggered part of the Frasnian/Famennian crisis. A single impact cannot explain the temporal and biological data given earlier, however. If the Frasnian/Famennian extinction is to be explained entirely by an extraterrestrial causation, then multiple impacts are clearly necessary (McGhee 1982:498; see also report of Kerr 1993a:176).

To return to the astonishing range of time duration estimates cited at the beginning of the chapter, we can now rule most of them out. This is not to say that the original estimates were unfounded or simply speculation. I have tried to show in this chapter how our understanding of the Frasnian/Famennian crisis has evolved with time. Earlier estimates have been refined with further data acquisition. This is how science progresses. The single key factor in our current understanding of the crisis is the vast improvement in time resolution that we now have, which was not available to earlier workers.

To summarize the findings of this chapter, the Frasnian/Famennian crisis occurs over 3.0 Ma at the maximum (table 3.2). The main pulse of the extinction spans only 2.0 Ma, from the Late *rhenana* Zone to the Middle *triangularis* Zone. The most devastating diversity loss, triggered by both high extinction rates and severe reductions in origination, occurs in the *linguiformis* Zone, which is 0.5 Ma or less in duration. At subzonal levels of time resolution this most severe drop in diversity did not span the entire *linguiformis* Zone, but only the last part of the zone. Several separate events are discernible within the later part of this zone, however, but all occur within one hundred thousand years or less.

IV

Victims and Survivors

A VERY PROMINENT paleobiologist once described lists of victims
and survivors as boring. I suspect some readers are in agreement,
and those will probably skip on to chapter 5. Yet how can one hope
to penetrate the mystery of what caused the Late Devonian mass
extinction without considering the catastrophic change that occurred
in the Earth's biosphere? I entreat the reader to persevere.

The Devonian is a fascinating time in the history of life on Earth.
Even the recognition of the Period by the great geologists of the last
century as a distinct phase in the geologic timescale makes compelling
reading (Rudwick 1979, 1985). It is during the Devonian that the
full-scale invasion of land occurs, the emergence of animal life from
the oceans, the dawn of the terrestrial ecosystem (figure 4.1). As land-
dwelling animals, we tend to think of "life" as other land-dwelling
animals and plants. Prior to the Devonian, however, there was no
terrestrial ecosystem to speak of. Some primitive plants, precariously
establishing a beachhead in protected coastal areas, was about it. The
interiors of the continents of the planet Earth were as barren as the
rocky landscapes of Mars.

Figure 4.1. Paleoecologic reconstruction of a Late Devonian landscape. Two primitive amphibians (*Ichthyostega*) are shown on land in the lower right. The trees are the club moss *Cyclostigma*. Ferns at the base of the nearest tree, and in the lower left foreground, are *Archaeopteris*. The creeper on the tree stump and fallen log at lower right is *Sphaenophyllum*. (From Spinar and Burian 1972. Used with permission of Mrs. Eva Hochmanová-Buri-ánová.)

Life, in the form of the earliest primitive bacteria, is present on the Earth some 3.5 Ga ago. That is the age of our oldest yet discovered moneran fossils, found in marine strata. Animal life, the kingdom of the multicellular heterotrophs, arose only some 600 Ma ago. And animal life remained in the protective arms of the ocean for two hundred million years. Things were getting a bit crowded in the oceans by the dawn of the Devonian, however. Food was scarce for

Figure 4.2. Paleoecologic reconstruction of a Late Devonian river. The two center animals are the primitive amphibians *Ichthyostega*. Freshwater fish are (left) the lobe-finned *Eusthenopteron* and (right) the lungfish *Rhynchodipterus*. Log in the river is the tree fern *Eospermatopteris*. (Artwork by Gregory Paul. From Exploring Earth and Life Through Time by Steven M. Stanley. Copyright (c) 1993 by W. H. Freeman and Company. Used with permission.)

some species, predators were abundant for others. The gradual spread of plants onto the land had set the stage: Food was now present out of the water (figure 4.1). Predators were absent. Animal life began to brave this harsh new environment in the Devonian—first the arthropods, then, in the Late Devonian, our ancestors (figure 4.2). They did not meet a gentle welcome.

What was the state of life on Earth before, and after, the Frasnian/Famennian crisis?

IN THE OCEANS: THE BOTTOM DWELLERS

Corals

One of the major global phases in the evolution of reef faunas occurs during the Devonian Period. Huge reef tracts, like the Great Barrier Reef in the modern world but an order of magnitude larger in extent, existed in the Devonian world in Australia, in western North America, across Europe, and in Russia. These ancient reefs would have seemed alien to us today, for unlike modern reef communities, Devonian reefs are dominated by odd tabulate corals and even stranger stromatoporoids. Both of those groups are extinct today. Similar to modern reefs, however, these ancient reefal ecosystems provided haven and habitat to an astonishing diversity of other marine species. (See figure 4.3 for a reconstruction of a Devonian off-reef community, a small part of the total complexity of the reefal ecosystem.)

Global reef ecosystems were decimated in the Late Devonian extinction event. The most severe biological crisis in the entire evolutionary history of both the ancient tabulate (see figure 4.3b, c, d, and f, and figure 4.4i) and rugosan (see figure 4.4b) corals occurs in this event, in terms of magnitude of diversity of species lost. Even though both coral groups survive the Frasnian/Famennian extinction, only to succumb to extinction in the great end-Permian mass extinction some 120 or so million years later, the numbers of species that perish in that final event are fewer than those that do not survive the Frasnian/Famennian crisis.

Figure 4.3. Paleoecologic reconstruction of a Middle Devonian off-reef community. a = swimming nautiloid cephalopod, b = tabulate coral colony (*Thamnopora*), c = tabulate coral polyps (*Syringopora*), d = tabulate coral individual, e = articulate brachiopod shellfish (*Athyris*), f = tabulate coral colony (*Syringopora*), g = stromatoporoid sponge. (From McKerrow 1978. Artwork by Elizabeth Winson. Copyright (c) 1978 by the Massachusetts Institute of Technology Press. Used with permission.)

The two coral groups do respond differently to the extinction event. The tabulate corals never really recover from the massive loss of diversity that they suffer in the Late Devonian, and they survive on in the later Paleozoic at diversity levels generally only a third of their previous numbers (figure 4.5). The rugosans, however, do manage to

Figure 4.4. Paleoecologic reconstruction of a Middle Devonian muddy shelf community. a = crinoid "sea lilies," b = rugose coral individual (*Mesophyllum*), c = bryozoan colony (*Fenestella*), d = walking trilobite arthropod (*Phacops*), e = articulate brachiopod shellfish (*Spirifer*), f = articulate brachiopod shellfish (*Athyris*), g = articulate brachiopod shellfish (*Kayseria*), h = articulate brachiopod shellfish (*Chonetes*), i = tabulate coral colony (*Heliolites*). (From McKerrow 1978. Artwork by Elizabeth Winson. Copyright (c) 1978 by the Massachusetts Institute of Technology Press, used with permission.)

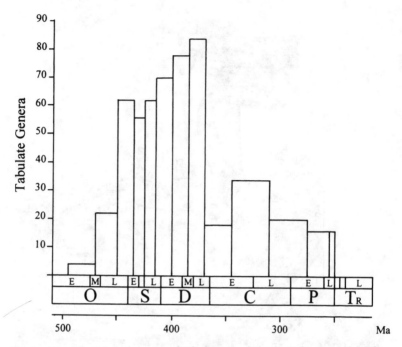

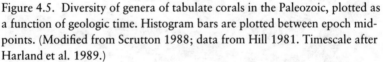

Figure 4.5. Diversity of genera of tabulate corals in the Paleozoic, plotted as a function of geologic time. Histogram bars are plotted between epoch midpoints. (Modified from Scrutton 1988; data from Hill 1981. Timescale after Harland et al. 1989.)

rebound almost to previous diversity levels in the later Carboniferous (figure 4.6).

Approximately 80% of all genera of tabulates are believed to perish in the extinction event. Scrutton (1988), however, notes that perhaps as few as seven genera survive into the Carboniferous, which would bring the estimate up to 92% extinction of all genera. Scrutton (1988) also links the failure of recovery of the tabulates to the ecology of the stromatoporoids, which are also decimated in the Late Devonian (see later), as the stromatoporoids are seen as creating many of the reefal niches within which the tabulates flourished.

The tabulates are exclusively colonial, and 57% of the Devonian fauna (excluding auloporoids; Scrutton 1988) was comprised of massive colonial growth forms. Massive cerioid perforate colonial forms (e.g., favositids) are most heavily decimated in the crisis, and although branching forms of the coenenchymal imperforate type are totally extinguished, they were of relatively low diversity prior to the event. In contrast, the rugosan coral group is comprised of both solitary and colonial forms. Interestingly, the solitary, nondissepimentate forms are virtually unaffected by the Frasnian/Famennian event, at least at the generic level (Scrutton 1988; see also Bambach 1985). Small, solitary individuals that lack dissepiments are some of the most primitive of the Rugosans, and generally inhabit deeper-water regions (a point to which we return in chapter 6).

More than one hundred genera of the chiefly colonial and larger solitary dissepimented forms go extinct in the event, representing a 60% loss in generic diversity (figure 4.6), according to Hill (1981). However, Hill's data are resolved only at the generic level of classification and at the epoch level of time. The more refined stage level data of Oliver and Pedder (1994) show that 84% of the Old World Realm rugosan genera and 100% of the New World Realm genera do not survive the Frasnian. (The "Old World" and "New World"

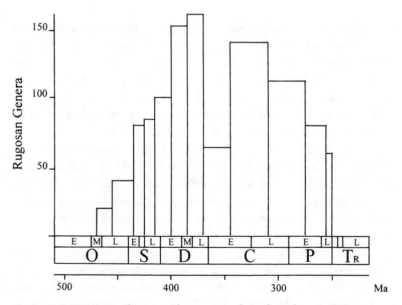

Figure 4.6. Diversity of genera of rugose corals in the Paleozoic. Histogram bars are plotted between epoch midpoints. (Modified from Scrutton 1988; data from Hill 1981. Timescale after Harland et al. 1989.)

biogeographic realms of corals will be discussed in more detail in chapter 5.) The even more refined substage data of Pedder (1982) and Sorauf and Pedder (1986) indicate that, in shallow-water marine realms at least, a species kill of 96% occurs, with only six species surviving the late Frasnian *linguiformis* Zone.

Scrutton (1988) points out that virtually no early Famennian corals are known and that they are assumed to exist somewhere chiefly because of the reappearance of some previously existent genera later in the Famennian (the "Lazarus Effect"). Only a few small solitary species reappear in the Early *rhomboidea* to Early *marginifera* zones in Poland (Rozkowska 1969, 1981). The environments inhabited by these survivors are interpreted to be of calm but turbid water, muddy and poorly aerated (Rozkowska 1969, 1981). Their tolerance for such environments may well indicate why they survived the Frasnian/ Famennian crisis.

Stromatoporoids

The other major faunal element in the great Devonian reefs are the stromatoporoids (see figure 4.3g). Stromatoporoids are a peculiar group of extinct sponges that secreted massive skeletons of calcium carbonate. Some paleontologists do not even believe they were sponges and prefer to consider them as an unusual group of corals.

The stromatoporoids lose some 46%, or roughly half, of their genera in the late Frasnian (figure 4.7). Stearn (1987) points out that, even though they were severely affected, the stromatoporoids fare much better than the corals (see also Webb 1994). In addition, they do not lose their reef-building potential, and although post-Frasnian stromatoporoid constructions are of small dimensions, they recover to construct reefs large enough to provide modern oil fields in Famennian strata of western Canada (Stearn 1987).

Late Famennian survivors generally are of the more primitive stromatoporoid orders, the clathrodictyids and the labechiids, the latter of which normally are only a minor part of the stromatoporoid fauna since the close of the Ordovician. The dominant Frasnian orders, in contrast, persist only in the survival of a single genus of each through the Frasnian/Famennian crisis.

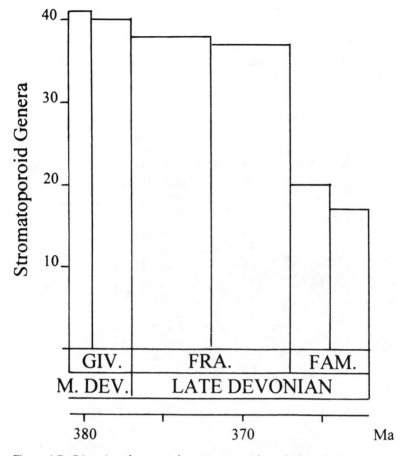

Figure 4.7. Diversity of genera of stromatoporoids in the later half of the Devonian Period, plotted as a function of geologic time. (Data from Stearn 1979, 1987. Timescale after Harland et al. 1989.)

Brachiopods

The brachiopods, largely unknown to most people today, are the dominant form of shellfish in the Devonian (see figure 4.3e, 4.4e-h). Walking the beaches of today's oceans, one commonly sees the numerous shells and shell fragments of the many molluscs that constitute most of the modern marine shellfish. Walking the beaches of the

Devonian world, the comparable shells and shell fragments were almost exclusively brachiopod.

The brachiopods lose more than 75% of their genera in the Frasnian/Famennian crisis (figure 4.8). The entire brachiopod orders of the Pentamerida and Atrypida go extinct, as they lose their

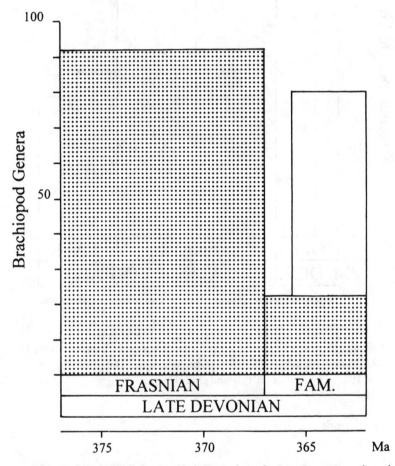

Figure 4.8. Diversity of genera of brachiopods in the Late Devonian, plotted as a function of geologic time. The shaded portion of the histogram shows the number of Frasnian genera that survive into the Famennian; the unshaped portion shows the number of new genera that originate later in the Famennian. (Data from Boucot 1975. Timescale after Harland et al. 1989.)

last species in the event. The orders of the Orthida and Strophomenida suffer significant familial losses. Within the Strophomenida, however, the Productellidae particularly survive to proliferate in the Famennian. The Spiriferida and Rhynchonellida likewise survive to constitute much of the Famennian brachiopod fauna (Johnson 1979). The brachiopods, considered as whole, make a diversity recovery to near previous levels in the later Famennian (figure 4.8), the timing of which is variable from region to region.

Foraminiferids

Foraminiferids are unicellular organisms of the kingdom Protista. They are very common in modern oceans, living both on the sea bottom as epibenthos and floating in upper water levels as epiplankton. Brasier (1988) notes that the first major radiation of the benthic foraminiferids comes during the Middle Devonian, rather late in the Paleozoic for a group whose evolutionary history extends back into the Cambrian. This first diversification and expansion of the foraminiferids continues into the Frasnian but is brought to an end during the Frasnian/Famennian crisis.

During the Frasnian, the foraminiferids form a distinctive microfauna in the tropical Paleotethyan region known as the Semitextulariina–Nodosarioid assemblage, an assemblage that is particularly decimated in the extinction event (Toomey and Mamet 1979; Poyarkov 1979). Some thirty species of the Semitextulariidae and Paratextulariidae perish as these two families become extinct during the diversity crisis (Brasier 1988), which also exterminates the last species of the families Multiseptidae and Nanicellidae. Although these four families did not survive, some fourteen others do, though barely for some (figure 4.9). The species of the Tournayellidae, for example, decline by more than 70% in the late Frasnian but reradiate in the later Famennian (Brasier 1988).

Brasier (1988) has documented a particular selectivity of extinction in the foraminiferids: The species with advanced septate architectures in their skeletal tests are eliminated. Species that do survive have primitive test architectures: simple spheres, uniserial tests with simple

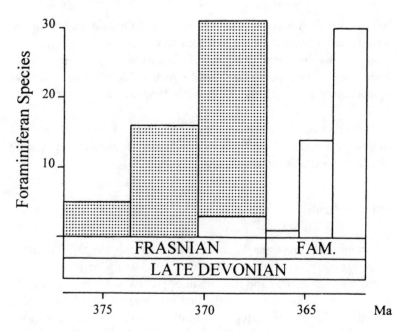

Figure 4.9. Diversity of species of foraminifera in the Late Devonian. The shaded portion of the histogram shows the number of species of the families Semitextulariidae and Paratextulariidae; the unshaded portion shows the number of species of the Tournayellidae. (Modified from Brasier 1988. Timescale after Harland et al. 1989.)

chambers, and simple agglutinated forms. Species with somewhat more advanced, or intermediate, pseudoseptate to quasi-septate test architectures also survive into the Famennian but suffer drastic reductions in their numbers (as with the species of the Tournayellidae). Skeletal test composition is also an important factor, in that all families having siliceous agglutinated tests survive, whereas almost half (45%) of those having calcareous tests do not.

Bryozoans

The bryozoans are tiny colonial organisms and, as lophophorate suspension feeders, are related to the previously considered brachiopods (see figure 4.4c). The colonial bryozoans, however, do not seem to suffer quite as severely as their cousins, the solitary brachiopods, during the Frasnian/Famennian crisis (Cuffey and McKinney 1979; Horowitz and Pachut 1993).

The diversification trend exhibited by the bryozoans for most of the Devonian is brought to a halt in the Late Devonian, with the loss of some 33% of their previous generic numbers (figure 4.10). Locally,

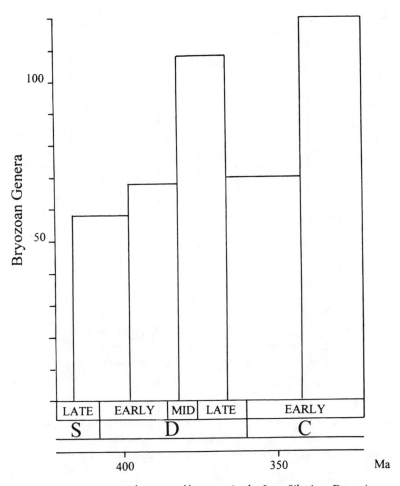

Figure 4.10. Diversity of genera of bryozoa in the Late Silurian, Devonian, and Early Carboniferous. (Data from Taylor and Larwood 1988. Timescale after Harland et al. 1989.)

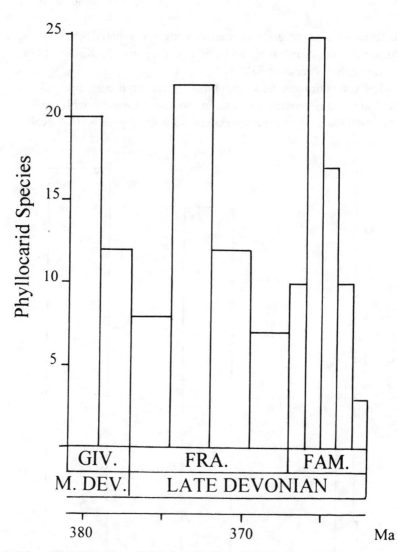

Figure 4.11. Diversity of species of phyllocarids in the Givetian and Late Devonian. (Data from Rolfe and Edwards 1979. Note also the short durations of the Famennian substage units compared to the Givetian and Frasnian, resulting from the short duration estimate of the Famennian in the timescale of Harland et al. 1989.)

the severity of species loss may be greater (Taylor and Larwood 1988). The effect of the crisis on the bryozoans is relatively slight at higher taxonomic levels, in that McKinney (1985) calculates that only three out of thirty-two families of bryozoans actually become extinct during the crisis.

Bigey (1986) notes the widespread disappearance of bryozoan faunas in western Europe and eastern North America just before the end of the Frasnian. Elsewhere, the bryozoans persist into the Famennian and are common in shelf facies in the Paleotethys region of the world (Troitskaya 1968). Finally, the group quickly recovers to previous diversity levels in the Early Carboniferous (figure 4.10).

Phyllocarids

The phyllocarids are malacostracan crustaceans, related to modern shrimp, lobsters, and crayfish. They are important members of marine ecosystems in the Devonian, along with other marine arthropod groups such as trilobites and ostracodes. The Middle and Late Devonian is punctuated by several crises for the phyllocarids (figure 4.11). Phyllocarids lose 68% of their species in the late Frasnian, and only seven species survive the very latest Frasnian (Rolfe and Edwards 1979). They recover and diversify in the early to middle Famennian, only to undergo an even more severe crisis in the late Famennian (figure 4.11), where 88% of phyllocarid species are lost, and only three species survive into the Early Carboniferous (Rolfe and Edwards 1979). Significant diversity losses also occur in the late Givetian to early Frasnian interval (figure 4.11).

Trilobites

The trilobites are an extinct group of marine arthropods, which, however, are very numerous in the Paleozoic and are important members of the vagile benthos (see figure 4.4d). The trilobites exhibit a steady decline throughout the Middle and Late Devonian (Alberti 1979; Briggs et al. 1988). Trilobites lose subfamilies throughout the Frasnian, and although the loss in standing diversity at the Frasnian/

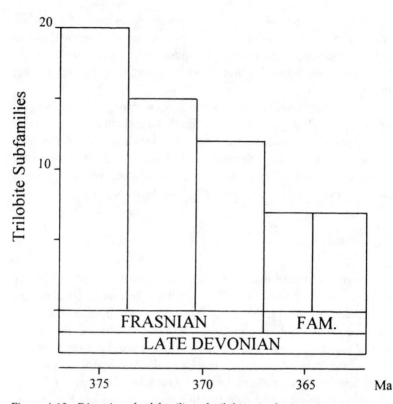

Figure 4.12. Diversity of subfamilies of trilobites in the Late Devonian. (Data from Briggs et al. 1988. Timescale after Harland et al. 1989.)

Famennian boundary (42%) is substantial, it is only slightly larger than diversity losses that occur within the Frasnian (figure 4.12).

Briggs and colleagues (1988) note, however, that those trilobites that do go extinct in the latest Frasnian belong to especially important groups, such as the family Scutellidae. Other extinctions appear to be geographically diachronous, in that the proetids, brachymetopids, and phacopids survive into the Famennian in Europe but virtually disappear in North America during the Frasnian. Even in Europe, however, local losses in trilobite diversity can be quite high (Feist and Schindler 1994).

The Frasnian/Famennian crisis appears to halt the previously steady decline in diversity, in that further decline does not occur within the Famennian (figure 4.12).

Ostracodes

Another major group of marine arthropods in Devonian oceans consists of the ostracodes, which are small bivalved crustaceans that have members in both benthic and planktic ecosystems. Benthic ostracodes, interestingly, exhibit a pattern of steady diversity decline during the Middle and Late Devonian (figure 4.13) that is similar to that

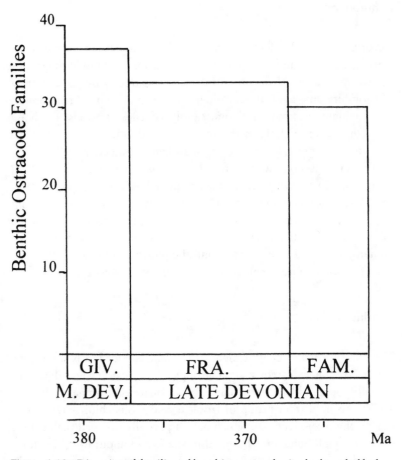

Figure 4.13. Diversity of families of benthic ostracodes in the later half of the Devonian Period. (Data from Gooday and Becker 1979. Timescale after Harland et al. 1989.)

exhibited by their relatives, the trilobites (figure 4.12). Overall, the benthic group loses four families in the Givetian and three in the Frasnian. As such, the Frasnian/Famennian crisis appears to be just another in a series of diversity losses that the group suffers during the span of the Middle Devonian to the Carboniferous (Casier 1989). Locally, however, benthic ostracodes have been shown to lose up to 65% of their species numbers at the Frasnian/Famennian boundary (Lethiers and Feist 1991).

Echinoderms

According to Paul (1988), data are very incomplete at present concerning survival and extinction of the various echinoderm groups across the Frasnian/Famennian interval. Apparently only one family of rhombiferans, the Callocystitidae, goes extinct, with three families of fissiculate blastoids and one family of edrioasteroids surviving through the interval. Thus the Blastozoa and Echinozoa do not appear to be very affected by the Frasnian/Famennian crisis.

The same is not the case for the Crinozoa (see figure 4.4a) and Asterozoa. Forty-two percent of asterozoan diversity is lost: Five families of asterozoans go extinct in the event, with only seven surviving into the Carboniferous (Sepkoski 1982b). Fifteen families, or some 32% of total familial diversity, of crinozoans go extinct in the Late Devonian crisis (McKinney 1985), and the paucity of Famennian crinozoan faunas is notable worldwide (McIntosh and Macurda 1979).

Benthic Algae

The fossil record of benthic algae as a whole is poor but is somewhat better for the calcareous algal groups. Some calcareous algae are actually rock builders, particularly in carbonate environments (Riding 1977). An important group of tropical calcareous chlorophytes consists of the receptaculitids, which have a partially calcified thallus that is generally globular to cylindrical in shape and about 6–15 cm long dimension (Tappan 1980). The calcareous chlorophytes are severely affected by the Frasnian/Famennian crisis. An entire major group, the Receptaculiteae, that originated in the Ordovician and flourished thereafter did not survive the Late Devonian (Nitecki 1969).

Glass Sponges

Thus far we have been considering major groups of marine organisms who are victims of the Frasnian/Famennian crisis. In contrast, the nonstomatoporoid sponges—in particular, the hexactinellid or "glass" sponges—are spectacular survivors.

The pattern of survival of the hexactinellids across the Frasnian/ Famennian boundary is particularly well preserved in the Devonian strata of New York State, where it can be seen that the group actually diversifies across the boundary (figure 4.14). Thirty-seven species of hexactinellids suddenly appear in the latest Frasnian in shallow marine habitats in New York, decline slightly in numbers across the boundary, and recover to roughly equivalent numbers in the early Famennian. After the crisis interval they sharply decline in diversity and are found in low species numbers in the later Famennian (figure 4.14). Similar blooms of glass sponge faunas during the crisis interval are also reported from other areas around the globe, such as Poland (Racki 1990) and western Canada (Rigby 1979; Geldsetzer et al. 1993). Hexactinellids appear to be rare in Late Devonian strata of Australia, but abundant lithistid demosponges (which also have siliceous skeletons) are reported (Rigby 1979).

Last, the glass sponges exhibit an interesting reciprocal ecological response, in comparison to just about every other element of the benthic fauna, to the Frasnian/Famennian crisis. Whereas other groups lose species numbers in the late Frasnian, the sponges actually diversify, and when "normal" marine organisms begin to recover species numbers in the mid to late Famennian, the sponges drastically decline (McGhee 1982). This would indicate that whatever environmental change occurs during the crisis is actually favorable to the glass sponges, just as it is deleterious to so many other organisms. This peculiar ecological change, from highly diverse Frasnian ecosystems to "simplified" Famennian ecosystems with large numbers of species of glass sponges, may provide a valuable clue to the ultimate cause of the mass extinction.

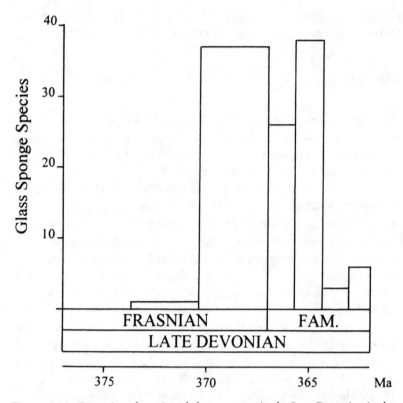

Figure 4.14. Diversity of species of glass sponges in the Late Devonian in the northern Appalachians (USA). (Modified from McGhee 1982. Timescale after Harland et al. 1989.)

Bivalves

Everyone today is familiar with the bivalve molluscs—the clams, scallops, oysters, and so on. They are ecologically diverse, being important elements of both the sessile epibenthos and vagile endobenthos. Some even swim and thus could be considered a minor part of the nekton. They are numerically abundant and constitute a large fraction of the modern shellfish fauna.

In the Devonian they were neither ecologically diverse nor numerically abundant. However, the bivalves are counted among the general survivors of the Frasnian/Famennian crisis. Although two families of bivalves, the Antipleuridae and Ambonychiidae, did become extinct during this interval of time, bivalve specialists conclude that the crisis "had a negligible effect on bivalves as a whole" (Hallam and Miller 1988). Most bivalve species are shallow-water dwellers during the Devonian, yet so also were many species of brachiopods that do not survive the crisis. Important, it is during the Frasnian that the first bivalve molluscs succeed in invading fresh-water regions of the terrestrial realm (Kriz 1979). We will return to this group when the terrestrial ecosystem is considered.

Gastropods

The gastropod molluscs, like their relatives the bivalves, are generally counted as survivors of the Frasnian/Famennian event. It is unclear at present, however, how much of this apparent lack of effect is due to inadequacies in the data, as this group is also generally not as well known as the bivalves. Sepkoski (1982b) lists only one family, the Palaeotrochidae, as going extinct in the Frasnian itself, and McKinney (1985) lists only four family extinctions (out of thirty-nine families total) for the Late Devonian. Linsley (1979) lists only eight genera that go extinct in the Frasnian. Interestingly, the gastropods also appear to have first invaded fresh, or at least brackish, water habitats during the Devonian (Dineley 1984).

Annelids

The annelids have a very fragmentary fossil record, as most have no skeletal tissues. Most are endobenthic, and much of what we know of their general evolution comes from the trace fossils of their burrows. The polychaetes have the best known skeletal fossil record, and they apparently suffer little in the Frasnian/Famennian crisis. Only two families, the Polychaeturidae and the Xianioprionidae, out of seventeen go extinct (Sepkoski 1982b).

IN THE OCEANS: THE SWIMMERS

Cephalopods

The major group of cephalopods present in the Late Devonian are the ammonoids, which first appear during the Early Devonian. Ammonoids appear superficially like the modern cephalopod *Nautilus,* but that similarity is indeed superficial (Ward 1992). The evolutionary history of this group is complex, and they experience a series of rapid increases and decreases in species diversity throughout the Devonian (House 1979, 1985; House and Kirchgasser 1993; Becker et al. 1989, 1993). They actually suffer five diversity crises during the Late Devonian alone (see table 3.1). The ammonoids lose diversity in the late Frasnian, and only eight genera survive the latest Frasnian (figure 4.15). With their characteristic high evolutionary rates, however, the ammonoids recover most of their generic numbers by the middle Famennian.

The related but less diverse nautiloid cephalopods (see figure 4.3a) also exhibit rapid evolutionary turnover in the Middle and Late Devonian, and they are affected by the Frasnian/Famennian mass extinction. Twenty-nine genera go extinct in the Frasnian, with only twenty surviving into the Famennian (Teichert et al. 1979). The nautiloids then rapidly rediversify in the Famennian, producing a total of forty-six new genera.

Conodonts

The conodonts are much smaller (in size) members of the epipelagic and mesopelagic nekton than the ammonoids and nautiloids. Conodonts were tiny eellike swimmers, having lengths of only about 4 cm. They are at present believed to be a type of chordate, and thus are related to us. They are totally extinct.

Aldridge (1988) notes that the Devonian "nadir of conodont diversity" is reached by these animals in the Frasnian—a nadir, however, that is short-lived. Almost all previous species of the important palmatolepids, ancyrognathids, and polygnathids disappear in the late Frasnian (Klapper and Ziegler 1979; Ziegler and Lane 1987). The

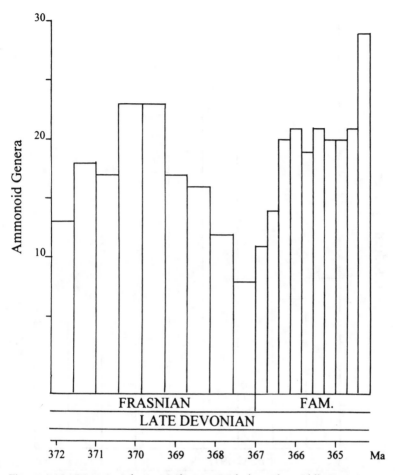

Figure 4.15. Diversity of genera of ammonoids from the middle Frasnian to the middle Famennian. (Data from Becker 1986. Timescale after Harland et al. 1989.) Notice the compression of ammonoid stratigraphic intervals in the Famennian relative to those of the Frasnian. Ammonoids probably evolved no faster in the Famennian than in the Frasnian. The apparent temporal shortening of Famennian stratigraphic units is an artifact of the Harland et al. (1989) timescale.

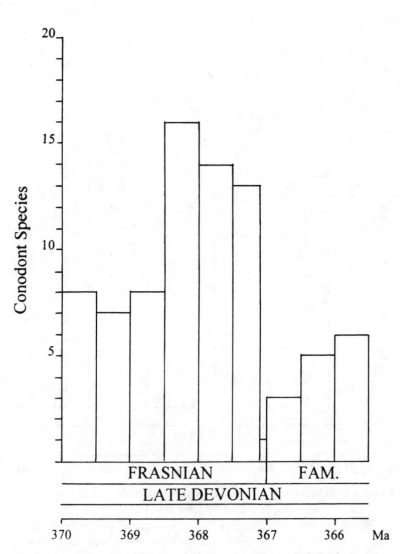

Figure 4.16. Diversity of species of the conodont genera *Palmatolepis, Meso-taxis,* and *Klapperina* in the last six conodont zones of the Frasnian and the first three zones of the Famennian. Each zone is estimated to extend 500,000 years (Ziegler and Sandberg 1990). In the latest Frasnian *linguiformis* Zone only a single species survives an extinction event, with an estimated duration of 20,000 years at the very most. (Data from Ziegler and Sandberg 1990. Age of the Frasnian/Famennian boundary after Harland et al. 1989.)

last species of *Ancyrodella* and *Ozarkodina* die. The last of the conodonts with true coniform apparatuses, so abundant in the earlier
Paleozoic, also disappear. In all, Aldridge (1988) estimates that about
89% of Frasnian conodont species do not survive into the Famennian.
The great majority of pelagic conodonts are individuals of the
species of the two genera *Palmatolepis* and *Polygnathus*. Only one
species of *Palmatolepis* (*P. praetriangularis*) and two species of *Polygnathus* (*P. brevilaminus* and *P. cf. planirostratus*) survive the latest
Frasnian (Sandberg et al. 1988). The biomass of neritic conodonts is
largely made up of individuals belonging to species of *Icriodus*. Only
two species of *Icriodus* (*I. alternatus* and *I. iowaensis*) survive the
Frasnian/Famennian extinction event (Sandberg et al. 1988).

The pattern of late Frasnian diversification and abrupt extinction
is particularly well exhibited by the important zonal conodonts (figure
4.16). Within the latest Frasnian *linguiformis* Zone, species diversity
drops from thirteen to one, during a geologic interval of time that
Sandberg and colleagues (1988) argue could not possibly be longer
than 20,000 years (see chapter 3). For a minimum estimate of the
time involved in the conodont crisis, Sandberg and colleagues (1988)
propose a few days at most.

Fishes

The only major members of the chordate group of animals present in
the Late Devonian world, excluding the peculiar conodonts, are the
fishes (see the appendix). Our amphibian ancestors, descendants of
these fishes, are present in the Frasnian world but in very low numbers, having just recently evolved (see later).

The fishes are severely affected in the Late Devonian mass extinction. All previously known fossil members of the agnathan (jawless)
fishes perish (figure 4.17). Westoll (1979) and Halstead (1965, 1966,
1988) point out that the agnathans had been on an evolutionary
decline since the earliest Devonian and that many of the groups that
perish in the Frasnian/Famennian crisis are represented by very few
species. However, Halstead (1988) also notes that two major agnathan groups, the heterostracans in Euramerica and the thelodonts
in Gondwana, flourish in the Frasnian. They do not survive into
the Famennian.

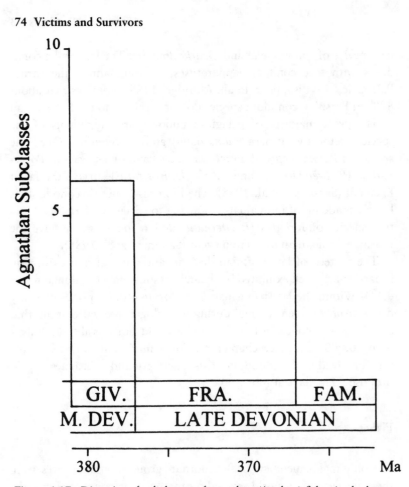

Figure 4.17. Diversity of subclasses of agnathan (jawless) fishes in the later half of the Devonian Period. (Data from Halstead 1988. Timescale after Harland et al. 1989.)

Living agnathan petromyzones (lampreys) are known from fossils as old as the Carboniferous, and living agnathan myxines (hagfish) have no fossil record. As the agnathans still exist on the Earth, some species must have survived the Frasnian/Famennian mass extinction somewhere.

The two principal groups of the Devonian gnathostome (jawed) fishes, the Placodermi and Acanthodii, survive the Frasnian/Famennian crisis but suffer major diversity losses. Fully half of the species present in the Frasnian do not survive into the Famennian (figure 4.18).

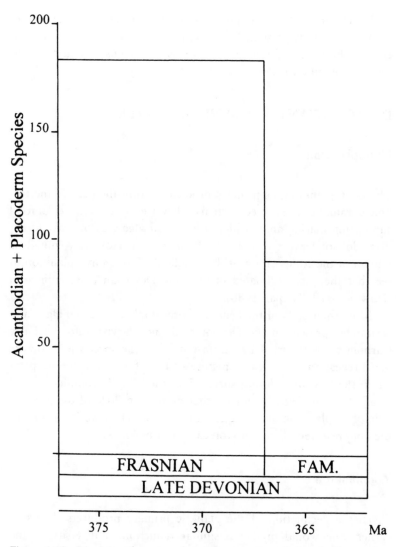

Figure 4.18. Diversity of species of gnathostome (jawed) acanthodian and placoderm fishes in the Late Devonian. (Data from Dennison 1978, 1979. Timescale after Harland et al. 1989.)

The chondrichthyans and osteichthyans are less numerous in Devonian seas than at present. Of these, however, two orders of marine chondrichthyans and four families of osteichthyans go extinct in the Late Devonian crisis (Sepkoski 1982b).

IN THE OCEANS: THE MINUTE DRIFTERS

Phytoplankton

The phytoplankton, as primary producers, form the base of the trophic pyramid of oceanic ecosystems. They are a critical source of food supporting marine animal life. Soft-tissued algae, as one would expect, do not leave the best of a fossil record. However, enough is known of the composition of the middle Paleozoic phytoplankton to see that they are decimated in the Late Devonian mass extinction (Downie 1979; Tappan 1980).

More than 60% of the genera of prasinophycean green algae and acritarchs, present in the Devonian, do not survive into the Early Carboniferous (figure 4.19). Data refined to the species level (figure 4.20) reveal an even bleaker picture—81% of acritarch species present in the Devonian do not survive into the Early Carboniferous. All in all, Tappan (1981) estimates that more than 90% of the preservable phytoplankton are affected in the event. The severity of loss in the nonpreserved phytoplankton can never be known.

Zooplankton

Given the decimation of the planktic primary producers discussed earlier, one would predict a similar reduction in diversity of the zooplankton, which is directly dependent upon the phytoplankton for food. However, the preserved record of the tiny animal plankton reveals a more complex picture than expected.

One important element of the Devonian zooplankton is the Cricoconarida, which are locally useful in biostratigraphy (Lütke 1979). These were peculiar, tiny, cone-shaped organisms, which are thought to be of molluscan affinity (see the appendix). Their mode

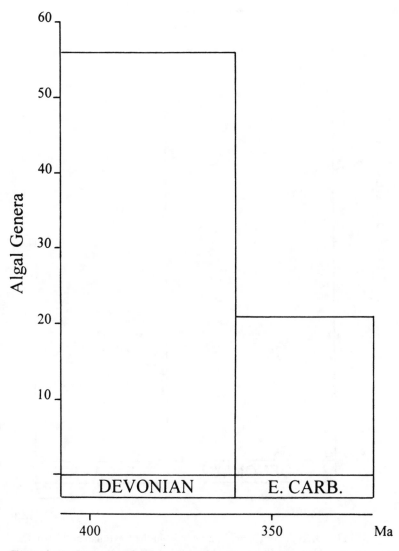

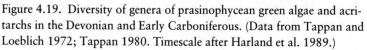

Figure 4.19. Diversity of genera of prasinophycean green algae and acritarchs in the Devonian and Early Carboniferous. (Data from Tappan and Loeblich 1972; Tappan 1980. Timescale after Harland et al. 1989.)

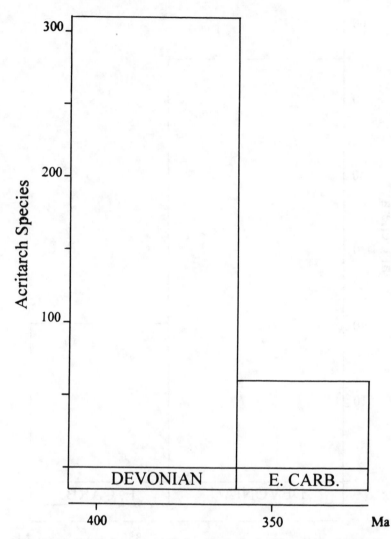

Figure 4.20. Diversity of species of acritarchs in the Devonian and Early Carboniferous. (Data from Tappan and Loeblich 1973; Tappan (1980). Timescale after Harland et al. 1989.)

of life may have been similar to modern planktic gastropods like the pteropods.

The cricoconarids are entirely extinguished in the Late Devonian crisis (figure 4.21). There is some confusion at present as to who actually survives and who does not. Species of only one family,

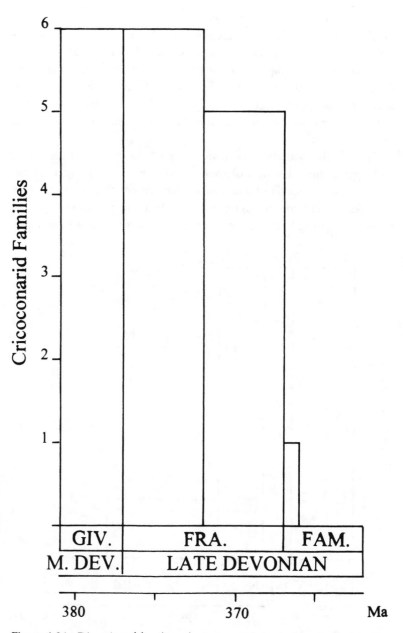

Figure 4.21. Diversity of families of cricoconarids in the later half of the Devonian Period. (Stratigraphic ranges from House 1975; Sepkoski 1982b; and Schindler 1990b. Timescale after Harland et al. 1989.)

the Styliolinidae, survive the Frasnian, and they quickly die out in the early Famennian, according to House (1975). Schindler (1990, 1993) also argues that only one family barely survives, and does not live any longer than the Early *triangularis* Zone of the very earliest Famennian (see figure 3.5), but that family is the Homoctenidae. Sepkoski (1992) agrees with House (1975) in having the styliolinids survive into the Famennian but has the homoctenids perishing in the Frasnian.

Though the overall familial diversity of the cricoconarids is not high, population densities of these organisms are very great. Thousands upon thousands of individuals of species of *Styliolina* and *Tentaculites* are commonly preserved upon bedding planes of Frasnian strata, indicating that great concentrations of these small enigmatic

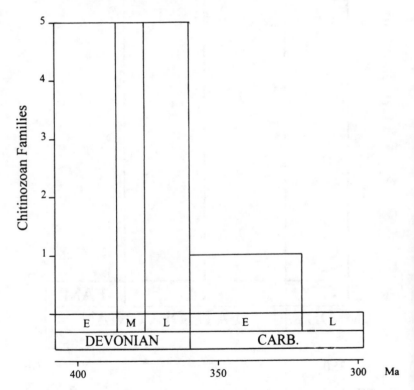

Figure 4.22. Diversity of families of chitinozoans in the Devonian and Carboniferous. (Data from Sepkoski 1982b. Timescale after Harland et al. 1989.)

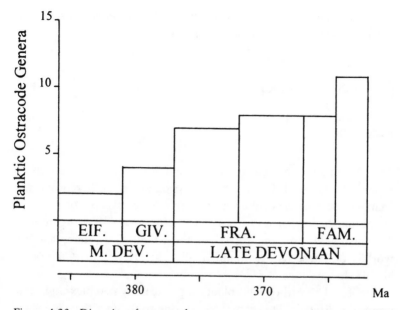

Figure 4.23. Diversity of genera of entomozoacean ostracodes in the Middle and Late Devonian. (Data from Groos-Uffenorde and Schindler 1990. Timescale after Harland et al. 1989.)

organisms filled the water column. In the Famennian they are gone, never to been seen again.

Another important, and even more enigmatic, group of Devonian zooplankton are the chitinozoans, which are thought to be of heterotrophic protistan affinity (see the appendix). These are tiny flask- or bottle-shaped organisms, which may occur singly or strung together in chains. Five families of these peculiar organisms appear in the fossil record in the Ordovician and range throughout the Silurian into the Devonian, where they are decimated (figure 4.22). No species of the Ancyrochitinidae, Conochitinidae, Hoegispharidae, or Lagenochitinidae survives the Late Devonian. Only one family, the Desmochitinidae, makes it into the Tournaisian, where it too dies out (Sepkoski 1982b).

Two last important groups of the preserved Devonian zooplankton are the entomozoacean ostracodes and the Radiolaria. Surprisingly, the tiny arthropods show no change in diversity at all across the Frasnian/Famennian boundary, and they even exhibit a sharp increase in numbers of genera in the late Famennian (figure 4.23). Similarly, the radiolarian protists apparently suffered no diversity losses in the Frasnian/Famennian crisis (Sepkoski 1982b).

FROM SEA TO LAND

Land Plants

The Devonian is a time of major significance in terrestrial plant evolution (Gray 1993). Major physiognomic and reproductive changes occur in the evolution of morphologies capable of surviving in the harsh thin air and dry environment of the land (figure 4.1). The first great forests evolved in the Middle Devonian, and the world's oldest seeds are found in the Late Devonian (Gillespie et al. 1981).

The fossil record of land plants is particularly complex and taxonomically difficult. This complexity is due to the fact that most land plants are preserved as fragments of individuals, such as leaves, roots, bark, or reproductive structures, rather than as a single complete organism. A separate taxonomy has evolved over time for each fossil aspect, such that we have "spore species" and "leaf species" described for different fragments of what may be a single individual.

Both microfossils and megafossils indicate that land plant diversity increases steadily during the Devonian, reaching a maximum during the Givetian (McGregor 1979; Boulter et al. 1988). Diversity then declines during the Late Devonian, but the pattern of that decline is not the same among the megafossils as among the microfossils. These data are puzzling. As both the microfossils and macrofossils are fragments of a single individual, one would expect changes in the diversity of plant individuals to be reflected by simultaneous changes in both of these separate fossil aspects of those individuals.

Spore species show a slight decline in numbers from the Givetian to the Frasnian, from fifty-seven species to fifty-one species. A dramatic drop in spore species occurs from the Frasnian to the Famennian (figure 4.24). Almost half (43%) of the species are lost, as diversity drops to twenty-nine species during the Famennian and remains at this level into the Tournaisian (Richardson and McGregor 1986). Streel (1992) argues that the end Frasnian dramatic drop in

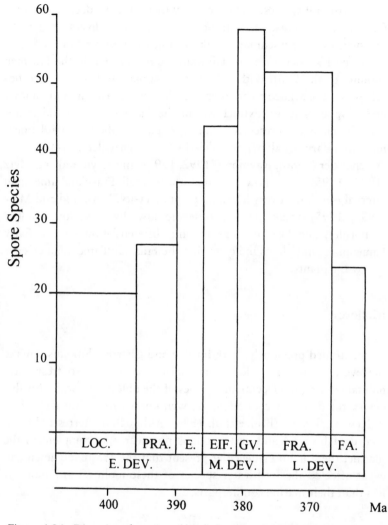

Figure 4.24. Diversity of species of land plant spores in the Devonian Period. (Modified from Boulter et al. 1988. Data from Richardson and McGregor 1986. Timescale after Harland et al. 1989.)

spore species diversity (figure 4.24) is a taphonomic artifact of poor preservation of spores in late Frasnian strata of New York State, which is the source area of the data reported by Richardson and McGregor (1986). However, Streel (1992) likewise maintains that a major loss in microfloral diversity does occur later in the Famennian.

Macrofloral species, however, show their greatest decline from the Givetian to the Frasnian, with no major change in diversity from the Frasnian to the Famennian (Banks 1980). Roughly half of the mega-fossil species listed from the Givetian are not present in the Frasnian (Boulter et al. 1988). Within the Late Devonian itself some macrofloral species are argued to go extinct in the early Frasnian, and another group appears to go extinct within the Famennian, but no major diversity change is presently seen to take place at the Frasnian/Famennian boundary (Chaloner and Sheerin 1979; Scheckler 1984).

Scheckler (1986), Raymond (1992, 1993), and Raymond and Metz (1992, 1995) note, however, that the overall Frasnian/Famennian interval is a time of very low land plant diversity. Raymond and Metz (1992, 1995) argue that the observed low diversity, low number of florules, and loss of biogeographic differentiation seen in Early Famennian land plants indicates that the Frasnian/Famennian crisis is felt by the plants.

Molluscs

As mentioned previously, both bivalve and gastropod molluscs make the evolutionary transition from marine ecosystems to freshwater habitats during the Devonian. Species of the oldest freshwater bivalve genus, the unioid *Archanodon,* are well known from Frasnian strata in New York State (Bridge et al. 1986). Although the terrestrial fossil record of bivalves and gastropods is not nearly as complete as the marine, there is no evidence at present that the Frasnian/Famennian event affected the freshwater species of these molluscs any more or less than their marine counterparts.

Arthropods

The arthropods successfully invade terrestrial freshwater and land environments in the Devonian, but the fossil record of the delicate myriapods and primitive hexapods is too fragmentary at present to reveal what effect, if any, the Late Devonian crisis had upon these early centipedes, insects, and kin (Shear et al. 1984).

The record for the predatory arachnids is somewhat better, particularly for the eurypterids (sea scorpions). The eurypterids lose 27% of their genera in the Frasnian/Famennian extinction (figure 4.25). This loss is small, however, in comparison to later Famennian extinctions, as fully 63% of the Famennian eurypterid genera do not survive into the Early Carboniferous (figure 4.25).

The small, bivalved, conchostracan branchiopods are generally found in freshwater to brackish-water regions. Conchostracans can be very abundant locally, but they are also very facies restricted and thus have a fragmentary fossil record (Rolfe and Edwards 1979). Novozhilov (1961) lists a total of thirty-three species for the early Frasnian, thirty species for the late Frasnian, but only one species for the Famennian. Tasch (1963) has questioned the taxonomy of Novozhilov (1961), however, and has cut in half the number of families recognized by him (Rolfe and Edwards 1979). Even if the number of Frasnian species recognized is inflated, it still appears that

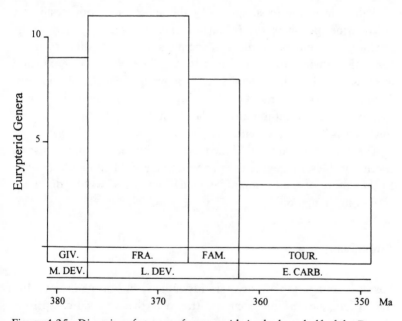

Figure 4.25. Diversity of genera of eurypterids in the later half of the Devonian Period and the earliest Carboniferous (Tournaisian Stage). (Data from Briggs et al. 1988; Plotnick 1983. Timescale after Harland et al. 1989.)

the conchostracans suffer significant diversity loss in the Frasnian/ Famennian crisis.

Fishes

The impact of the Frasnian/Famennian extinction on the fishes in total is outlined earlier. Important, however, is the fact that the fish had evolved many freshwater species in the Devonian. Fossils of freshwater agnathans, placoderms, acanthodians, and osteichthyans are known from the stratigraphic record. The last group is of particular importance, as it is within the class Osteichthyes (the "bony fishes," see the appendix) that we find the air-breathing fishes, fishes that have lungs, such as the chondrosteans, crossopterygians (lobe fin fish, see figure 4.2), and dipnoians (lungfish, see figure 4.2). The crossopterygian fish are ancestors to the first tetrapods, the amphibians, which evolved just before the Frasnian/Famennian crisis.

Although all the fossil species of the agnathans, both freshwater and marine, perish in the Frasnian/Famennian mass extinction, the freshwater members of the Placodermi and Acanthodii fare much better than their marine counterparts (McGhee 1982). The extinction event was ecologically selective in both groups: For the placoderms, 23% of the freshwater species die relative to a 65% kill of the marine species; for the acanthodians, 30% of the freshwater species go extinct versus 87% of the marine species.

The differential survival of the freshwater species versus the marine species of the same higher taxa of fishes may yield an important clue to the nature of the Frasnian/Famennian crisis and will be discussed further in chapter 6.

Amphibians

Fossils of our tetrapod ancestors first appear in Frasnian strata, but then a long gap in time occurs: "After a further interval comprising most of the Famennian, a more substantial sample of early tetrapod material, including complete skeletons, appears in late Famennian horizons" (Ahlberg and Milner 1994). During this time gap our

ancestors, the amphibian tetrapods, were present in such low numbers that none of their skeletons have yet been found. The tetrapods
had experienced the Frasnian/Famennian mass extinction.

The first tetrapod vertebrates had long been thought to evolve only
in the late Famennian (where their fossil remains are undisputable),
and it was believed that the vertebrate invasion of terrestrial realms
only begins at the virtual close of the Devonian. This would mean,
of course, that the Frasnian/Famennian crisis does not affect the
amphibians, simply because there are none present at the time. They
would have evolved after the crisis was over.

Recent findings have overturned this view, however, and it appears
that our ancient land ancestors were indeed unfortunate enough to
have experienced whatever triggered the Frasnian/Famennian crisis.
Trackways made by tetrapods have been found in Frasnian strata in
Australia (Warren and Wakefield 1972). Ahlberg (1991) reported the
first fragmentary skeletal remains of a Frasnian age tetrapod from
Scotland. Late Frasnian tetrapod remains are now known from Lat-
via, Scotland, and Australia (Lebedev and Clack 1993; Ahlberg and
Milner 1994). The newly discovered Famennian tetrapod *Hynerpeton
bassetti* from eastern North America extends the known geographic
range of early amphibians to a virtually global equatorial distribution,
suggesting an evolutionary history for the group that may extend as
far back as the Early Devonian (Daeschler et al. 1994).

Panderichthyid lobe-finned fishes are now also known to have been
much more tetrapodlike than previously thought (Ahlberg and Milner
1994). These lobe-finned fishes were present in the early Frasnian and
appear to have adopted a shallow water predatory niche. They have
dorsally placed eyes, useful for an animal lying low and concealed in
the water while searching the shore for arthropod prey. Imagine a
quiet stream flowing though the strange plants of the Late Devonian
landscape. An unwary arthropod strays close to the water's edge,
spray explodes as the panderichthyid bursts from the stream pro-
pelled by its stubby lobe fins, a quick snap of toothy jaws, and the
arthropod is lunch. It has even been suggested that the panderichthy-
ids were capable of extensive and prolonged terrestrial locomotion
(see Ahlberg and Milner 1994).

The transition from stubby lobe fins to stubby limbs with feet has
occurred by the middle to late Frasnian. The amphibians have

evolved, in most respects still anatomically fish with the exception of their limbs (see figure 4.2). They leave their footprints in the mud by the streams in Australia; the bones of another are buried in the sands in Scotland. Life is good. They are the sole vertebrates present in the new world, the terrestrial world (see figure 4.1). There are no predators to worry about, no competitors for the arthropod prey. Then comes the mystery of whatever happens in the latest Frasnian, and it is decidedly not good.

Obviously, with such fragmentary fossil material as has been found at present, it is not possible to judge the effect of the Frasnian/ Famennian crisis on the Class Amphibia. However, they are present in the Frasnian. Why do they reach population sizes large enough to ensure that at least some individuals are preserved in the fossil record again only in the latest Famennian? The long gap in time: the four- to five-million-year time gap between their first appearance in the late Frasnian and their reappearance in the late Famennian is suggestive. The Amphibia, our early land ancestors, may have been dealt a severe blow in their very infancy.

SUMMARY: "BREADTH" OF THE FRASNIAN/ FAMENNIAN CRISIS

The crisis certainly meets the geographic and ecologic "breadth" criterion of a mass extinction, as discussed in chapter 2. The entire planet is affected. From the bottoms of the oceans, from the surface of the seas, from the beaches, rivers, lakes, and land—species vanish everywhere.

Most major taxa are decimated: cnidarians, stromatoporoids, brachiopods, foraminiferids, phyllocarids, echinoderms, benthic algae, cephalopods, conodonts, fishes, phytoplankton, zooplankton. For some groups already suffering bad times, the bad times get really worse, as is seen in the record of the trilobites and ostracodes. Small and large, active and sessile, on the bottom and at the surface, marine life is particularly hard hit. Land animals and plants seem not to suffer as much, but this may be due to the more fragmentary nature of the fossil record of terrestrial habitats. Only the annelids, and their molluscan cousins the bivalves and gastropods seem to be relatively unaffected by the mass extinction.

Survivors of the decimation appear peculiar. They generally represent the "simplest" forms of previous evolution in the taxa involved, the survival of more primitive, more ancestral, morphologies. This is especially noted for the cnidarians, stromatoporoids, and foraminiferids.

One group, and one group only as far as I have been able to determine, actually diversifies when all other groups suffer severe diversity losses: the hexactinellid sponges. That is, they not only survive (or are neutrally not affected), but actually blossom while all else is dying. We will return to examine these animals in greater detail in chapter 6.

The catastrophic change in the Earth's biosphere that took place in the Frasnian/Famennian mass extinction can now be fully seen in the list of victims and survivors compiled in this chapter. This list is more than a mere compilation; it contains a signal, an ecological signal, that can tell us something about why all these animal and plant species vanished. However, before we can filter the ecological signal out of the data we must know more about the environmental context of the Late Devonian world, a world very different from our own. The nature of the Earth during the Late Devonian is the subject of the next chapter.

V

The Late Devonian World

THE DEVONIAN world was very different from ours. This we know for a certainty, but just how different has been the subject of considerable debate and disputation. We know the present positions of the continents are not fixed; they are moving even now with respect to one another, riding along as the gigantic plates of the Earth's crust and upper mantle grind inexorably past each other, expanding away from the hot molten rock upwelling in the oceanic ridges and plunging back into the mantle in the depths of the oceanic trenches, eventually to be remelted within the Earth. The ultimate recycling scenario.

What was the Late Devonian world like, the world that was to be stricken by the Frasnian/Famennian mass extinction? Where were the continents?

THE PROBLEM OF THE PRE-MESOZOIC WORLD

Unfortunately for us, the recycling of the Earth's plates imposes a time limit on easy, or at least fairly easy, paleogeographic reconstructions of past continental positions. The oldest ocean floor still preserved is early Mesozoic; the Paleozoic ocean floor is long gone—subducted into the Earth, metamorphosed, mangled, and remelted. Continental positions in the Mesozoic and Cenozoic can be accurately determined by referring to the ages of zones of the ocean floor, which are youngest at the midocean ridges and progressively older away from the ridges. We can even determine the rate at which continent A has moved away from continent B in the past, or project into the future using those same rates to reveal a world yet to come—a world in which southern California is no longer where it is now and in which the eastern part of Africa has split off and become a separate continent.

Things are more difficult in the Paleozoic. The ancient ocean floors are long gone. The only data we have concerning the positions of the continents in the Paleozoic exist on the continents themselves. Those data can be artificially divided into two groups: the paleomagnetic and the nonpaleomagnetic. I say "artificially" because one would hope that all data would be used together to come to the best fit for Paleozoic continental positions. Indeed, that is being attempted today, but it was not in the beginning, as we shall see.

EARLY BIOGEOGRAPHIC DATA

Many types of biogeographic data have been used in paleogeographic reconstructions, from plants to animals, from sea to land. By way of example, only two sources of data will be briefly discussed here. These are the data provided by the geographic distribution species assemblages of brachiopods and cnidarians. Both of these animal groups are marine, and they are sessile (i.e., they do not move around as adults). Their geographic range, over wide areas, depends upon dispersal of their larval stages by oceanic water currents.

The distribution of Emsian (Early Devonian) brachiopod provinces is given on a modern geographic base in figure 5.1. By far the largest province is that of the Old World. Species of the Old World Province extend across Europe, Africa, and Asia. They also jump the modern great oceans to occur far to the south in Australia and to the west in northern Canada and the western United States. In the rest of the

Americas, two provinces are present, the Appalachian and Malvino-kaffric. The Appalachian Province extends from eastern Canada and down the eastern United States; then it jumps to western South America in Columbia and Venezuela. A specialized cold-water brachiopod fauna is called the Malvinokaffric Province, and it extends from the Appalachian fauna (from which it evolved) down into Brazil and Argentina in South America, but it also jumps the Atlantic to occur in South Africa and Antarctica.

The distribution of Emsian brachiopod provinces obviously makes no sense in a modern geographic context. Entirely different provinces are but a short distance away from one another in some parts of the world, whereas in other parts species of the same province occur widely separated from each other.

There is nothing unusual about the Emsian brachiopods. Other organisms also show nonsensical geographic distributions, when plot-

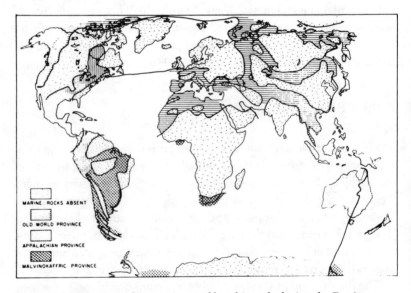

Figure 5.1. Biogeographic provinces of brachiopods during the Emsian (Early Devonian), plotted on modern world geography. Note the continuation of the Appalachian Province from eastern North America to western South America and its extension, the cold-water Malvinokaffric Province, into central and southern South America. (From Johnson and Boucot 1973. Used with permission.)

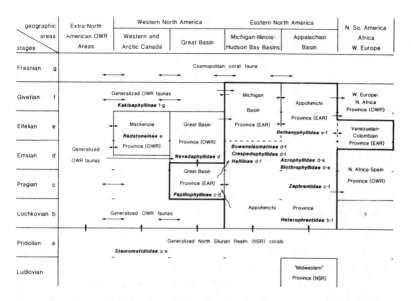

geographic areas stages	Extra-North American OWR Areas	Western North America			Eastern North America		N. So. America Africa W. Europe
		Western and Arctic Canada	Great Basin	Michigan-Illinois-Hudson Bay Basins	Appalachian Basin	
Frasnian g				Cosmopolitan coral fauna			
Givetian f		Generalized OWR faunas *Kakisaphyllinae* f-g		Michigan Basin Province (EAR)	Appohimchi Province (EAR)	W. Europe-N. Africa Province (OWR)
Eifelian e	Generalized OWR faunas	Mackenzie *Redstoneinae* e Province (OWR)	Great Basin Province (OWR) *Nevadaphyllidae* d	Great Basin Province (EAR) *Bowenelasmatinae* d-f *Craspedophyllidae* d-f *Halliinae* d-f	*Bethanyphyllidae* e-f *Acrophyllidae* d-e *Blothrophyllidae* d-e	Venezuelan-Colombian Province (EAR)
Emsian d						
Pragian c			Great Basin Province (EAR) *Papiliophyllinae* c-d		*Zaphrentidae* c-f	N. Africa-Spain Province (OWR)
Lochkovian b		Generalized OWR faunas		Appohimchi *Heterophrentidae* b-f	Province	?
Pridolian a		Generalized North Silurian Realm (NSR) corals *Stauromatidiidae* a-e				
Ludlovian				'Midwestern' Province (NSR)		

Figure 5.2. Biogeographic provinces of rugose cnidarians during the Late Silurian and Devonian. The faunal provinces of the Eastern Americas Realm (EAR) are given within the heavily lined box below the Eastern North America geographic column. Note the extension of Eastern Americas Realm faunas into northern South America (Venezuelan–Columbian Province) in the Emsian and Eifelian. (From Oliver and Pedder 1989. Artwork courtesy of W. A. Oliver, Jr. Reproduced with the permission of the Association of Australasian Palaeontologists.)

ted on modern continental positions. The rugose cnidarians have a particularly interesting history, evolving from widespread cosmopolitan species to narrowly adapted local endemic species, back to widespread cosmopolitan species during the Late Silurian and Devonian (Oliver and Pedder 1989, 1994; Oliver 1990). In the Devonian, cnidarian faunas are divided into two large biogeographic units: the Old World Realm (OWR) and the Eastern Americas Realm (EAR). Each realm is subdivided into a series of smaller provincial units (figure 5.2). The Old World Realm cnidarians have a geographic range very similar to the one seen above for the Old World Province brachiopods during the Early and Middle Devonian, jumping from Europe to North America to Africa. The Eastern Americas Realm cnidarians have a distribution similar to that of the Appalachian Province brachiopods, jumping from North to South America during the same period of time (figure 5.2).

These geographic distributions, as mentioned earlier, make no biological sense when plotted on modern geographies. Obviously, the continents and oceans were in entirely different places during the Devonian than they are today. But where were they?

THE ENTRANCE OF PALEOMAGNETIC DATA

Following the revolution in the geological sciences produced by plate tectonic theory in the 1960s and 1970s, the rush was on to reconstruct the ancient geographies of the past. The Paleozoic was a problem, as previously discussed. On the surface, however, there should be no problem at all. Continental rocks, as well as oceanic ones, are capable of taking on the magnetic field of the Earth when they are deposited. Therefore, it should be possible simply to measure the orientation of the magnetic field in a continental rock sample to determine the ancient pole positions, and the declination of the field to determine the latitudinal position. And *voilà!*—you have your ancient continental position. As you would guess, it turns out not to be so simple. The magnetic signal in the sample may not be the true original field orientation; the true signal may have been diagenetically altered, overprinted, or erased. Unfortunately, geologic processes that might lead to such diagenetic alterations are extremely common on the continental land masses.

Several paleogeographic reconstructions of the Devonian world were produced during the middle and late 1970s, although paleomagnetic data were still sparse for many areas of the world (Smith et al. 1973; McElhinny 1973; Morel and Irving 1978; Scotese et al. 1979; Bambach et al. 1981). Some are very detailed, others less so. I will choose the detailed reconstruction by Scotese and colleagues (1979) for discussion here (figure 5.3); it can also be seen in a very nice color representation in Bambach and colleagues (1981).

North America, northern and eastern Europe, and Greenland are shown together as parts of a single continent, called Laurussia by Scotese and colleagues (1979) and Bambach and colleagues (1981) but now more commonly named Euramerica. Euramerica is posi-

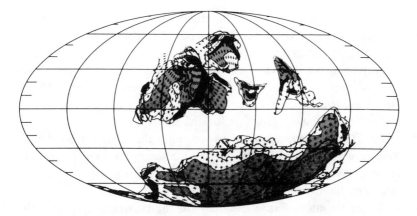

Figure 5.3. Paleomagnetic reconstruction of the Emsian world according to Scotese et al. (1979) and Bambach et al. (1981). Note the large ocean, some 3,000 km wide, separating Euramerica on the equator from Gondwana in the Southern Hemisphere. Deep oceans are unshaded in the figure. Heavy shading indicates mountainous regions; medium shading, dry land; and light shading, shallow marine waters. [From the *Journal of Geology* (Scotese et al. 1979). Published by the University of Chicago. Copyright (c) 1979 by the University of Chicago Press, used with permission.]

tioned so that the equator falls in the Great Lakes region of North America and between Denmark and Sweden in Europe (figure 5.3). Siberia is a separate continent, located to the northeast of Euramerica between 30°N and 60°N. The region around Kazakhstan is also shown as a separate continent, aptly named Kazakhstania, and is just to the east of Euramerica and near the equator. China and Malaysia are shown as a single continent just to the east of Kazakhstania.

The rest of the continents of the world are shown as part of a single large supercontinent named Gondwana and located in the Southern Hemisphere (figure 5.3). It includes the modern continents of South America, Africa, Australia, and Antarctica, as well as southern Europe, the Near East, and India. The south pole of the Earth is located in east central Africa, and the eastern margin of Gondwana (i.e., Australia) is up near the equator. The maps also show active tectonic zones (regions of mountain building), extent of coverage of the continents by shallow seas, and geographic distribution of the land surface.

This map (figure 5.3) and the others mentioned earlier are based on paleomagnetic data. Bambach and colleagues (1981) do show other paleogeographic data as well, however, such as the distribution of evaporites (xeric environmental indicators) and coal swamps (mesic environmental indicators). And some anomalies appear. In the Devonian, evaporites occur in Siberia, indicating hot and dry conditions. Yet Siberia is shown as located above 30°N in the high latitudes. The discrepancies are even worse for the Early Carboniferous (not shown here), where coal swamps are shown to occur extensively above 30°N, even as high as 60°N, in the paleomagnetic-based reconstruction (Bambach et al. 1981).

By far the most glaring anomaly (to the biostratigraphers) is the very large ocean shown to exist between southeastern Euramerica and northwestern Gondwana (figure 5.3). A huge oceanic gap, over 3,000 km wide, is shown between two areas of the world that have shallow marine faunas of the same biogeographic province. How could the sessile animals of the Appalachian marine communities have maintained reproductive contact with members of their species in South American marine communities, particularly with an oceanic gap in between, which would have had strong westward-directed water currents? The tiny larvae, passive floaters in the plankton, would have been swept to the west in both regions; they could not possibly have crossed such a large oceanic gap. Maintenance of reproductive contact would have been impossible. With genetic continuity broken, the populations would evolve in separate directions; they would become separate species.

Something was very wrong here. The biogeographic data said that the same species existed in southeastern Euramerica and northwestern Gondwana in the Emsian (figure 5.1). The paleomagnetic data said a huge ocean existed between the two regions in the Emsian (figure 5.3). Both could not be right.

THE ENTRANCE OF NONPALEOMAGNETIC DATA

Because we have noticed what we believe are *serious palaeoclimatic incompatibilities with palaeomagnetic positionings of certain Devonian continents,* we have . . . reconstructed probable positions of the Devonian continents based only on lithic and tectonic data, and determined that fossil distributions are related in a reasonable fashion.

Heckel and Witzke 1979:99; italics mine

And Heckel and Witzke amassed a truly impressive amount of nonpaleomagnetic data: global geographic occurrences of carbonates, biogenic, nonbiogenic, and oolitic; the evaporites, halite, potash, and sulfates; phosphorites and bauxites; the silicate clastics, redbeds, and black shales; cherts; coals. All these sediments yield valuable paleoclimatic data concerning absolute temperatures and seasonal ranges in temperature, average rainfall in mesic areas and degree of aridity in xeric areas, degrees of oceanic circulation from open to restricted to stagnant. To these data they added tectonic information, where the mountains were forming in the Devonian, valuable clues of continent–continent interactions. And finally, they factored in the biogeographic data of generations of paleontologists: the geographic distributions of cnidarians, stromatoporoids, calcareous algae, ammonoids, and brachiopods. It was a monumental effort, and it produced a very interesting paleogeographic reconstruction (figure 5.4).

Perhaps the most noticeable difference between figure 5.4 and figure 5.3 is the absence of the large ocean between Euramerica and Gondwana. The position of Gondwana has actually not changed at all; it is the other continents that have moved, and they have all moved south. The equator is now up in northwestern Canada in Euramerica, and not in the Great Lakes. Siberia is now mostly below 30°N and near the equator. Some other changes are apparent, such as the breakup of China and Malaysia into several continental fragments, but these are not germane to the discussion at hand.

The biogeographic anomaly of figure 5.3 has been largely resolved in figure 5.4. Southeastern North America is in direct contact with northwestern South America. A continuous shallow sea connects the two regions with their shared species and communities. The south equatorial current sweeps marine larvae from Europe westward across northern Canada into the western United States, aptly explaining the continuation of the European Old World Province species in those two areas of Euramerica (see figure 5.1).

If this nonpaleomagnetic reconstruction is correct, then the paleo-

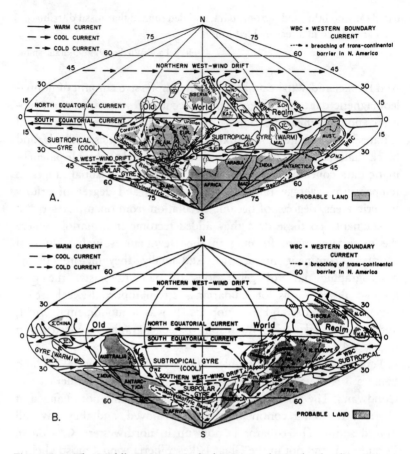

Figure 5.4. The Middle Devonian paleobiogeographic and paleoclimatic reconstruction of Heckel and Witzke (1979). Note the positioning of southeastern North America against northwestern South America at about 45°S, the equator in Alaska in western Euramerica and in northern Australia in eastern Gondwana, and the south pole in central Africa. (From Heckel and Witzke 1979. Artwork courtesy of P. H. Heckel. Reproduced with the permission of the Palaeontological Association.)

magnetic data are somehow wrong. Leaving aside for the moment the problem of sample diagenesis, Heckel and Witzke (1979) note that paleomagnetic reconstructions rest on three assumptions: dipolar field, geocentricity, and the statistically averaged coincidence of magnetic poles and rotational axis of the Earth. They maintain that the sedimentologic and biological data most accurately reflect atmospheric and oceanic circulation patterns, which are determined by the rotation of the Earth; hence these data best represent the rotational axis of the planet. If the paleomagnetic data do not agree, argue Heckel and Witzke (1979), then the Devonian magnetic poles did not correspond to the rotational poles of the planet.

The magnetic poles do not coincide with the rotational poles of the Earth today, so is this conclusion any surprise? Yes, say paleomagnetic specialists. The magnetic poles wander about with time; they fluctuate about the rotational poles. These oscillations occur in human time frames, but in geologic timescales they should average out and be coincident with the rotational poles. Heckel and Witzke (1979) suggest that this might not be true, and that the magnetic poles may have deviated from the rotational poles for significant periods of time in the Devonian.

THE ULTIMATE NONPALEOMAGNETIC RECONSTRUCTION: PALEOZOIC PANGAEA?

Early Devonian paleogeographies based largely on interpretations of paleomagnetic data *fail to satisfy both biogeographic and climatic indicators.* . . . The pangaeic reconstruction suggested herein relies chiefly on paleobiogeographic data and on varied climatic indicators.

Boucot and Gray 1983:580; italics mine

What is a pangaea? Pangaea is the "all-earth," the huge supercontinent that existed at the end of the Paleozoic and beginning of the Mesozoic and that began to break up some 200 Ma ago, eventually to produce the modern continental distributions. This supercontinent was a single landmass, a continuous contiguous continent composed of all the separated landmasses that we know as individual continents today.

Paleomagnetic data indicate that Pangaea did not exist until the latest Paleozoic, although its biggest component, Gondwana, did. (See the preceding discussion of continents and continental fragments included in Gondwana.) The other landmasses of the Earth were separate continents. Slowly, the other landmasses of the Earth coalesced in phases of continental accretion throughout the Paleozoic until the northern continents finally collided with Gondwana to produce Pangaea, the "all-earth" continent, in the Permian.

The Devonian Period occurs at about the halfway point in this accretion process. Euramerica (itself composed of two previously separated continents, Baltica and Laurentia), Siberia, Kazakhstania, and so on, are still separated continents in the Devonian (figures 5.3, 5.4). Or are they?

Boucot and Gray (1983) and Boucot (1988) suggest that they might not have been. The problem of the pre-Mesozoic world again arises. No one disputes the existence of Pangaea at the beginning of the Mesozoic, and its subsequent fragmentation during the Mesozoic and Cenozoic. The history of that fragmentation can be traced out on the ocean floors of the Earth. But where is the oceanic evidence for all the proposed shifting about, collision, and coalescence of separated continents during the Paleozoic? Not from the ocean floor; none is preserved. The evidence comes from paleomagnetic data from the continents themselves.

Boucot and Gray call those data into question in a much more sweeping fashion than Heckel and Witzke (1979). They suggest that Pangaea existed for the entire Paleozoic, and not just at the end. It did not form late in the Paleozoic; it was always there. And Boucot and Gray (1983) and Boucot (1988) gather nonpaleomagnetic data to support that reconstruction.

Of particular interest at this point is that they give a reconstruction of the world during the critical time in question: the Late Devonian Epoch. Note that in this reconstruction all the continental landmasses of the Earth are shown in the Southern Hemisphere, below the equator (figure 5.5). The south pole sits in southern Africa; the equator just touches Alaska, Siberia, and Australia/New Guinea.

The similar marine communities of southeastern North America and northwestern South America are not positioned as closely as in the Heckel and Witzke (1979) reconstruction, but there is no oceanic gap with through-flowing water currents. Thus continuity within the Appalachian biological province is maintained.

Boucot and Gray (1983) indicate that the Late Devonian world was one of low climatic gradient, in comparison with the Early Devonian. That is, the temperature gradient from the pole to the equator was not as steep during the Late Devonian. The equatorial region is not as hot; the polar region is not as cold. The latter point is of

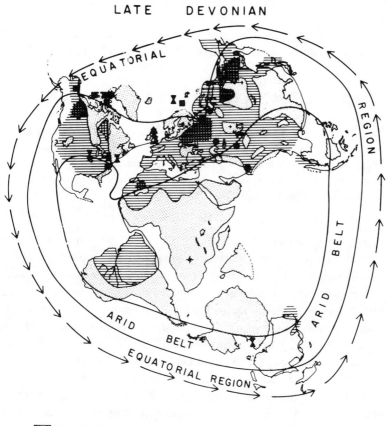

LATE DEVONIAN

Land & Areas Lacking Early
Late Devonian Strata

Marine

Marine Evaporites

+ South Polar Region

● Laterite Products Other
Than Bauxite & Kaolin

↗ Inferred Surface Current Direction

■ Coal

I Calcretes

▲ Aegerine - Bearing
Lacustrine Sediments

● Bauxite Occurrences

Figure 5.5. A pangaeic reconstruction of the Late Devonian world, where all the continents are assembled together and all are placed in the Southern Hemisphere of the Earth (below the equator) after the work of Boucot (1988) and Boucot and Gray (1983). (From Boucot (1988) in the *Canadian Society of Petroleum Geologists Memoirs*, vol. 14(3), pp. 211–227. Used with permission.)

considerable climatic significance, and we will return to in chapter 7, when we consider Late Devonian glaciation scenarios.

Was this the geographic state of the world that was to be struck with the Frasnian/Famennian mass extinction? A Devonian Pangaea?

THE RETURN OF PALEOMAGNETIC RECONSTRUCTION

Two conflicting models have been proposed for the middle Paleozoic position of Gondwana with respect to Euramerica. A pole from Morocco . . . suggests that a broad ocean (~3,000 km) separated the northern continents from Gondwana, whereas results from Mauritania . . . indicate that this ocean had closed by Devonian time.

Hurley and Van der Voo 1987:138; italics mine

Hurley and Van der Voo pinpoint the major area of discrepancy between nonpaleomagnetic and paleomagnetic reconstructions of the Devonian world, but they do so only when the "closed-ocean" model had gained support from paleomagnetic data. Kent and colleagues (1984) had suggested a repositioning of Gondwana based on a new paleomagnetic pole determination from the Devonian Gneiguira Supergroup in Mauritania, northwestern Africa. This paleopole determination suggests that northwestern Gondwana is located much further to the north, and thus much closer to Euramerica, in the Late Devonian. The new orientation, modeled by Scotese (1984), is illustrated in figure 5.6A. A variant of the previous paleomagnetic reconstruction for the Emsian (figure 5.3) is given for the Late Devonian in figure 5.6B for contrast. In figure 5.6B the contentious 3,000 km ocean exists between Euramerica and Gondwana; in figure 5.6A it does not.

The key difference in the two paleomagnetic versions in figure 5.6 is the positioning of Gondwana; Euramerica stays in the same position in both. This is in direct contrast to the solution suggested by the nonpaleomagnetic reconstructions (figures 5.4 and 5.5), both of which keep Gondwana in the same position but move Euramerica further south. In figure 5.6A the disputed 3,000 km ocean is closed by bringing the northwestern margin of Gondwana up near the equator, so that eastern North America is placed in proximity to northwestern Africa. The consequence of this rotation is that the eastern margin of Gondwana is further south, moving northern Australia to

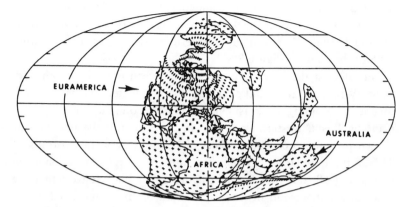

LATE DEVONIAN (A)

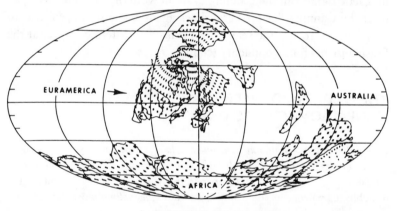

LATE DEVONIAN (B)

Figure 5.6. Two contrasting paleogeographic reconstructions of the Late Devonian world, which largely differ in how Gondwana is pivoted relative to Euramerica (which remains in the same position). In reconstruction (A) the eastern margin of Gondwana (Australia) is rotated down to about 30°S latitude, in reconstruction (B) it is rotated up to about 10°S latitude. (From Hurley and Van der Voo 1987. Used with permission of the authors.)

about 30°S. In the previous disputed paleomagnetic reconstruction (with the 3,000 km ocean) the northwestern margin of Gondwana is around latitude 30°S and the northern coast of Australia is within 10°S of the equator (figure 5.6B).

But at least the anomalous 3,000 km ocean is finally eliminated, right? Wrong! Hurley and Van der Voo conducted extensive paleomagnetic sampling of Frasnian/Famennian boundary strata in the Canning Basin, western Australia, in order to test the two models shown in figure 5.6. Their results placed northwestern Australia at 15°S in the Late Devonian and thus support the reconstruction given in figure 5.6B, not that given in figure 5.6A. They suggest that the Gneiguira Supergroup in Mauritania had been remagnetized in the late Paleozoic, and thus the paleopole determined by Kent and colleagues (1984) cannot be used for a Late Devonian reconstruction.

Thus the contentious 3,000 km ocean still survives, saved by paleomagnetic data from the Canning Basin in western Australia. We shall meet the Canning Basin several times again in the future, for it is to play a major role in the search for evidence of asteroid impact at the Frasnian/Famennian boundary (chapter 9).

THE WEDDING OF PALEOMAGNETIC AND NONPALEOMAGNETIC DATA?

New paleomagnetic data have become available in the past 5 yr that require modifications in previously published paleogeographic reconstructions for the Silurian and Devonian. In this paper, the new paleopoles are compared to published paleogeographic models *based on paleoclimatologic and biogeographic data.*

Van der Voo 1988; italics mine

Two radically different new reconstructions are proposed by Van der Voo (1988). One is for the Early Devonian and shows a Devonian world in which the 3,000 km ocean no longer exists between Euramerica and Gondwana (figure 5.7). Then, just when you think this matter has finally been laid to rest, the ocean reappears again in the Late Devonian (figure 5.8). How could this be?

New paleomagnetic data for the Early Devonian shift Euramerica substantially south of the latitudes shown in figures 5.6A and 5.6B. Rather than an equator that falls in the Great Lakes, the equator now

Figure 5.7. The Early Devonian paleomagnetic reconstruction of Van der Voo (1988). Note the absence of the large ocean seen in figure 5.3, and the positioning of eastern Euramerica against western Africa, and the south pole in Argentina. Explanation of biogeographic codes (letters in circles): A = Eastern Americas Realm (including brachiopod Appalachian Province), M = brachiopod Malvinokaffric Province, R = Old World Realm (Rhenish–Bohemian Province). (From Van der Voo 1988. Used with permission of the author.)

is positioned through southern Greenland, upper Hudson Bay, and northern British Columbia in Canada (figure 5.7). Most of Euramerica is thus now positioned south of the equator. They also shift northwestern Gondwana somewhat further to the north, so that eastern North America is now in contact with northwestern Africa.

Van der Voo (1988) notes that this reconstruction solves a series of tectonic and biogeographic problems, which had led to the rejec-

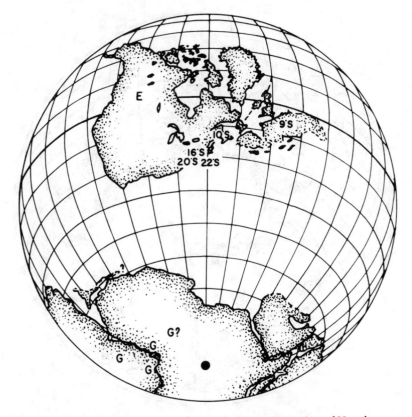

Figure 5.8. The Late Devonian paleomagnetic reconstruction of Van der Voo (1988). Note the appearance of a large ocean between Euramerica and Gondwana, resulting from the positioning of the south pole in central Africa in Gondwana, and the equator in Hudson Bay in Euramerica. (From Van der Voo 1988. Used with permission of the author.)

tion of previous paleomagnetic reconstructions by many Devonian stratigraphers and paleontologists. Eastern Euramerica is now seen to collide with western Gondwana in the Early Devonian, thus providing a plate tectonic explanation for the Acadian orogeny (mountain-building-period), which is known to occur during this time. Similarly, the known provincial distribution of marine communities makes sense in the new reconstruction, and Van der Voo (1988) plots these biogeographic (i.e., nonpaleomagnetic) data on his paleomagnetic-derived reconstructions (figure 5.7).

In many respects, the Van der Voo (1988) paleomagnetic Early Devonian reconstruction in figure 5.7 is very similar to the Heckel and Witzke (1979) nonpaleomagnetic Middle Devonian reconstruction in figure 5.4. Ignoring the locations of Siberia and the continental fragments of China in the Heckel and Witzke (1979) reconstruction, the chief difference is seen in that Heckel and Witzke (1979) position southeastern North America against northwestern South America, whereas Van der Voo (1988) positions eastern North America against northwestern Africa. Might these be minor differences in paleogeographic fine tuning? Has a consensus been reached? Unfortunately, the answer is no.

In the Late Devonian reconstruction of Van der Voo (1988) the contentious ocean returns between Euramerica and Gondwana (figure 5.8). Why? New paleopole positions for Euramerica still suggest the newer southern positioning of the continent, and though the continent moves north a little, the equator is only slightly lower than that shown for figure 5.7. It is the western margin of Gondwana that has rapidly moved away from Euramerica. The south pole is shown to occur in southern Argentina in the Early Devonian (figure 5.7) but in central Africa in the Late Devonian (figure 5.8), requiring, of course, a large rotation of Gondwana to the south. The data supporting this paleopole position are the previously discussed Frasnian/Famennian boundary samples from the Canning Basin in western Australia. Van der Voo (1988) notes some nonpaleomagnetic data that seem to support such a Frasnian/Famennian boundary reconstruction—the occurrence of glacial sediments in South America and Africa (figure 5.8), which would be understandable given their close proximity to the reported south polar position. We shall see in chapter 7, however, that these glacial sediments occur in the Middle *praesulcata* Zone (see table 1.6 for the zonations), that is, just below the Devonian/Carboniferous boundary.

A LATE DEVONIAN EURAMERICA—GONDWANA SUPERCONTINENT?

The best-fitting paleolatitudinal positions for North America and South America suggested by the Andreas and Snowy River pole positions allow a

supercontinent configuration that is virtually identical to those proposed by
McKerrow and Ziegler (1972) and Keppie (1977), *which are based largely
on faunal distributions.*

Miller and Kent 1988:198; italics mine

Miller and Kent (1988) published new data from Pennsylvania, which
suggest that Euramerica is located much further south of the equator
than previously thought. They place the eastern margin of Euramerica
against northwestern Gondwana in the Late Silurian and Early Devo-

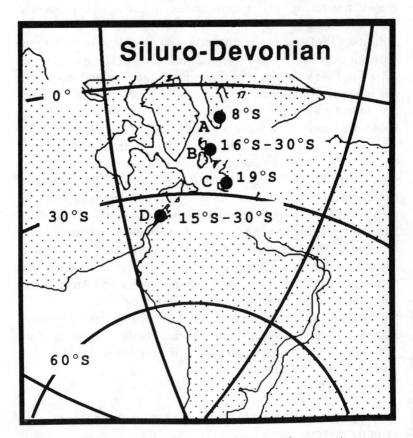

Figure 5.9. The Early Devonian (and Late Silurian) paleomagnetic reconstruction of Miller and Kent (1988) for Euramerica and western Gondwana.
Note the positioning of the eastern margin of Euramerica against the northwestern margin of South America. (From Miller and Kent 1988. Used with
permission of the authors.)

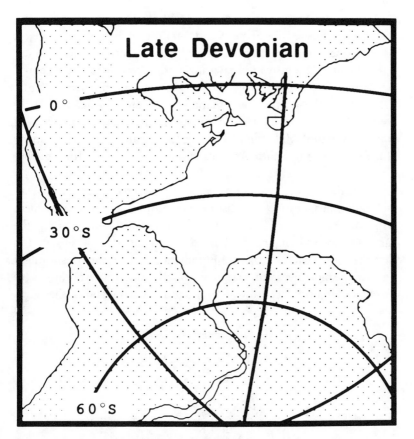

Figure 5.10. The Late Devonian paleomagnetic reconstruction of Miller and Kent (1988). Note the positioning of southeastern North America against northwestern South America at about 30°S, as Euramerica is shown to have moved northeasterward, right laterally, with respect to Gondwana from the Early to Late Devonian. No major ocean between Euramerica and Gondwana appears in this reconstruction. (From Miller and Kent 1988. Used with permission of the authors.)

nian (figure 5.9). More important for our considerations of the Late Devonian world, they maintain contact between Euramerica and Gondwana during this period. Euramerica is interpreted as moving to the northeast, grinding in a right-lateral fashion past Gondwana, so that the southern portion of Euramerica is against northwestern South America by the time of the Late Devonian (figure 5.10).

A new Famennian reconstruction, without the anomalous ocean, was also published by Scotese and McKerrow (1990). They reject paleomagnetic data placing the south pole in central Africa and use paleoclimatic data instead in favoring a pole position in north-central Argentina. The resulting "tight-fit" Famennian reconstruction (figure 5.11) places the northeastern margin of Euramerica against northwestern Africa in Gondwana. Their Euramerican positioning is very similar to that of Miller and Kent, but Gondwana is rotated counterclockwise, moving South America further south and Africa further north.

Last, the recent Frasnian paleogeographic reconstruction of Dalziel and colleagues (1994) positions Gondwana more in agreement with Miller and Kent (1988). Euramerica remains in more or less the same position as in figures 5.12 and 5.10, but Gondwana is rotated so that the southeastern margin of Euramerica (Appalachian) is brought against the northwestern margin of South America (Andean). This reconstruction is in accord with biogeographic data, and further gives

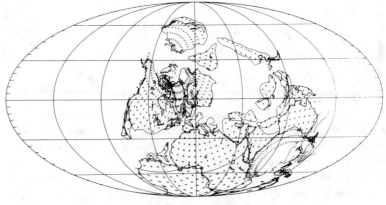

Late Devonian (Famennian)

Figure 5.11. The Late Devonian paleogeographic reconstruction of Scotese and McKerrow (1990). Note the positioning of southwestern Europe against northwestern Africa, the equator in Hudson Bay in Euramerica and against northern Australia in eastern Gondwana, and the south pole in Argentina in South America. No major ocean between Euramerica and Gondwana appears in this reconstruction. (From Scotese and McKerrow 1990. Reprinted with permission of the Geological Society Publishing House, Avon, England.)

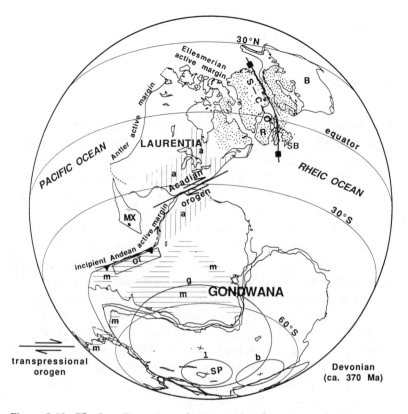

Figure 5.12. The Late Devonian paleogeographic reconstruction of Dalziel et al. (1994). Compare with figure 5.10 for the similar reconstruction of Miller and Kent (1988). Vertical lines and lowercase letters *a* indicate the geographic range of the brachiopod Appalachian Province, horizontal lines and lowercase letters *m* the Malvinokaffric Province. Stippling in northeastern Euramerica represents the extent of the redbeds of the "Old Red Sandstone Continent," sediments shed to the east and west of the mountains produced by the Scandian–Caledonian orogeny (S-C-O in the figure). (From Dalziel et al. 1994. Artwork courtesy of I. W. D. Dalziel.)

an intriguing plate tectonic explanation for the Acadian orogeny (figure 5.12). The Appalachian and Andean mountain systems are thus seen to be genetically related to one another, even though at present one is on the eastern margin of North America and the other is on the western margin of South America. The final collision of

Euramerica with the African margin of Gondwana is then proposed to occur only at the end of the Paleozoic.

SUMMARY: GOING FULL CIRCLE?

The debate concerning the Late Devonian world is by no means closed. Yet the debate does seem to have gone full circle: from early biogeographic data and stratigraphic data suggesting a close connection between eastern North America and western South America to paleomagnetic data suggesting a 3,000 km oceanic separation between the Americas, back to combined data sources once again suggesting a close fit of the two continental masses in the Late Devonian.

The reconstructions of Scotese and McKerrow (1990) (see figure 5.11) and Dalziel and colleagues (1994) (see figure 5.12) will be used in various discussions throughout the remainder of this book. Of immediate concern is the discussion of the ecological signature of the Frasnian/Famennian mass extinction to follow in chapter 6. The positioning of the continents and oceans during the critical Frasnian/Famennian period is very important, not a mere abstract academic exercise. If we are to tease out the ecological signature in the extinction and survival data given in chapter 4, we must know what "high latitude" is during the Late Devonian and where "low latitude" is located, and we must know how high is high and how low is low.

VI

Ecological Signature of the Mass Extinction

THE MASSIVE deterioration in ecosystems that occurred throughout the world during the Frasnian/Famennian crisis can correctly be described as catastrophic (McLaren 1982). Frasnian ecosystems are ecologically very diverse and equitable in structure. Early Famennian ecosystems, in contrast, are impoverished in ecological diversity and in overall species richness. Primitive or "simple" forms of life are common, forms with high tolerances to environmental stress, such as cyanobacteria, encrusting algae, and poriferans (particularly glass sponges). In some parts of the world reeflike mound structures, constructed by layers of cyanobacterial filaments and layers of algae, are found in early Famennian marine environments, sitting atop the dead cnidarian and stromatoporoid reefs of the Frasnian. Their presence may have more to do with the absence of grazing organisms than with their own evolutionary success. In normal times grazers would consume these cyanobacterial and algal layers before they could accumulate.

The effect of the biotic crisis can easily be seen in the "bottleneck" constriction of ecological diversity that occurs in the Appalachian region of eastern North America (figure 6.1). Frasnian benthic marine ecosystems are dominated by sessile filter-feeding brachiopods, which is normal for the Paleozoic world. They constitute only 30% of total species numbers, however, and the remaining 70% of the species represent sessile filter-feeding bivalves and predaceous cnidarians, vagile bivalve detritus feeders, sessile and vagile echinoderm filter feeders and predators, vagile gastropod grazers and predators, nektic cephalopod predators, and zooplanktic cricoconarids.

The previously diverse and ecologically complex Frasnian ecosystem is replaced by an ecologically depauperate Famennian ecosystem after the crisis. Species diversity is low in early Famennian marine environments, although population sizes of those species that do

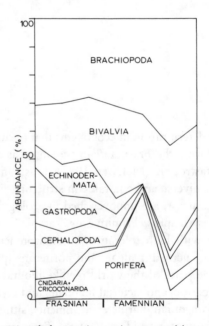

Figure 6.1. Proportional changes in species compositions of shallow marine ecosystems during the Frasnian and Famennian of New York State. Early Frasnian ecosystems support high diversities of organisms; early Famennian ecosystems are depauperate of diversity and are overdominated by poriferans (hexactinellids), brachiopods, and bivalves. (From McGhee 1982.)

exist may be quite large. The rapid proliferation of opportunistic individuals of survivor species is a common ecological phenomenon following a disturbed ecosystem. In eastern North America, early Famennian marine environments are proportionately overdominated by reduced species numbers of only three types of organisms: poriferans (particularly glass sponges), surviving brachiopods, and bivalve molluscs. The hexactinellid, or glass, sponges may constitute as much as 40% of local species numbers.

The analysis of the fossil remains of organisms around the globe that perish—as well as those that survive—during the Frasnian/Famennian interval of time (chapter 4) reveals interesting differential, and nonrandom, ecological effects of the mass extinction. Parts of the ecological signal can be detected locally, other parts only on a global scale. Using what we know of Late Devonian global paleogeography (chapter 5), let us begin with the large-scale signal.

LATITUDINAL EFFECTS

"The most extensive reef development this planet has seen"—that is the assessment of Copper (1994:7) concerning the great reef ecosystems that existed from the Middle Devonian into the Frasnian half of the Late Devonian. He estimates that these reef ecosystem tracts covered 5,000,000 km^2 of marine seaboard at their maximum development, almost ten times the areal extent of reefal ecosystems seen in the modern oceans. In the later Famennian, following the mass extinction, reef complexes covered less than 1,000 km^2 at the very most.

From these figures it can be clearly seen that tropical reefal and peri-reefal marine ecosystems are particularly hard hit during the Frasnian/Famennian biotic crisis. The low-latitude, geographically widespread, and massive stromatoporoid-tabulate reefal ecosystems vanish, and peri-reefal rugose coral-tabular stromatoporoid bioherms are decimated. For comparative ecology, however, these statements have little explanatory power. What is needed is a differential signal, not just a total loss. Such a signal may be provided by the stromatoporoids, brachiopods, foraminiferids, and trilobites.

Though the stromatoporoids suffer a severe reduction in biomass during the mass extinction, they do not become totally extinct, nor do they totally lose their reef-building potential (see chapter 4). Stearn (1987:678) stresses the ecological selectivity of the mass extinction: "An F/F [= Frasnian/Famennian] mechanism must explain the persistence of the clathrodictyids, *the resurgence of the labechiids (a minor group after Ordovician time)*, and the great decline in the actinostromatids, hermatostromatids, and stromatoporoids" (italics mine).

When the distribution of the survivors is plotted in the Late Devonian world geography, Stearn (1987) argues, a geographic pattern of selectivity emerges as well (figure 6.2). Stearn (1987) used the "closed-ocean" paleogeographic reconstruction of Scotese (1984), discussed

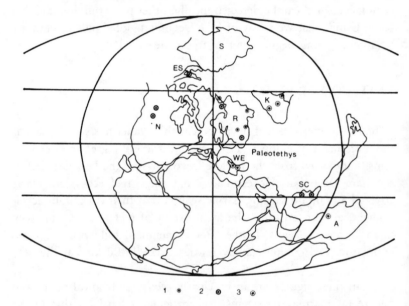

Figure 6.2. Geographic distribution of stromatoporoid faunas dominated by labechiids (cool-water species), and faunas where labechiids are absent (remnant Frasnian species), during the latest Famennian. Key for faunas: 1 (solid circles) faunas with labechiids only, 2 (open circles) mixed faunas, 3 (circles with star center) faunas without labechiids. Key for regions: N = North America, ES = eastern Siberia, S = Siberia, R = Russia, K = Kazakhstan, WE = western Europe, SC = southern China, A = Australia. Paleogeographic reconstruction used is that of Scotese (1984). See figure 5.6A of this book. (From Stearn 1987. Artwork courtesy of C. W. Stearn.)

in the preceding chapter (figure 5.6A). More modern reconstructions place Euramerica in a more southerly position, with the equator running through Hudson Bay (see figures 5.11 and 5.12), but the outcome of Stearn's reasoning is not significantly altered if his data are plotted in these newer configurations (but would be if they were plotted on one of the anomalous "open-ocean" reconstructions discussed in chapter 5).

Post-Frasnian stromatoporoids are of small dimensions and are generally found in the warm-water equatorial region of the Paleotethys (figure 6.2). The Paleotethys (sometimes also called the western Rheic Ocean; see figure 5.12) is an equatorial body of water bounding central Europe on the north, eastern North America on the west, and northern South America and Africa on the south. It is in this region that the remnant Frasnian species are found alone, without labechiids. In other parts of the world the relict Frasnian species occur with labechiids, or labechiids occur alone.

Relict Frasnian stromatoporoid faunas can occur as far north and south of the equator as 30° of latitude—that is, in shallow marine waters as far north as Kazakhstan or as far south as Australia (figure 6.2). Eastwardly flowing equatorial water masses are deflected into higher latitudes within the Paleotethys region, however. These deflected equatorial water masses warm the higher-latitude margins of the Paleotethys Ocean, much as the Gulf Stream keeps warm the high-latitude eastern margins of the Atlantic Ocean today (i.e., northern Europe, which otherwise would have the climate of northern Canada). Stearn (1987) notes that this distribution of relict Frasnian faunas, without labechiids, closely matches the geographic distribution of the warm-water benthic ostracode *Hypsela—Rectonaria* Province of Lethiers (1983).

Outside the warm water Paleotethys, in the north-facing seaways of the Russian Platform, the west-facing seas of western North America, and the high-latitude waters of eastern Siberia, the labechiids occur. Sometimes they are mixed with relict Frasnian forms; sometimes they occur alone (figure 6.2).

Stearn (1987) believes the labechiids to be better adapted to cool water conditions than the majority of Frasnian species, which are tropical and low latitude in distribution. If the Frasnian/Famennian crisis is triggered by global cooling, the resurgent spread of the labechiids would be understandable, as well as the survival of the relict Frasnian faunas in the Paleotethys.

Differential survival of high-latitude, cool-water-adapted species is also exhibited by the Brachiopoda (Copper 1977), which are the dominant form of shellfish in Frasnian benthic ecosystems. Of the total brachiopod fauna, approximately 86% of Frasnian genera do not survive into the Famennian. However, 91% of brachiopod families whose species are generally confined to low-latitude, tropical regions perish in the mass extinction (figure 6.3). In contrast, there is a loss of 27% of the brachiopod families with species members that range into high-latitude, cool-water regions of Gondwana (figure 6.3). The pattern is striking: virtual elimination of taxa with species confined to low latitudes, much less severe loss of taxa that have high-latitude species members.

Another element of the sessile benthos that exhibits latitudinal patterns of survival is the foraminiferids (Kalvoda 1986, 1990). The foraminiferids suffer major losses in species diversity, with substantial reduction in the global belt of carbonate sedimentation that occurs at

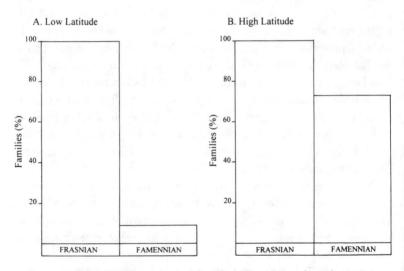

Figure 6.3. Proportional survival of families of brachiopods with species that inhabited low-latitude, tropical regions (A) and families with species that ranged into high-latitude, cool-water regions (B) in the Frasnian and Famennian. (Data from Copper 1977.)

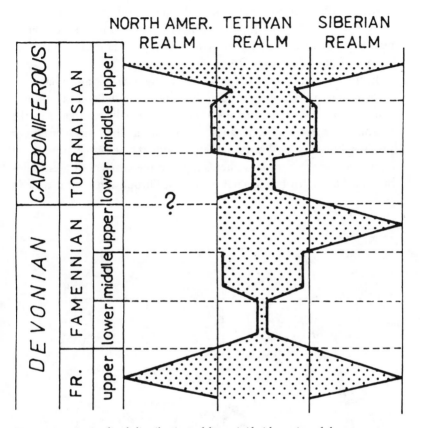

Figure 6.4. Latitudinal distribution of foraminiferid species of the warm-water, equatorial, Tethyan Realm during the Late Devonian. In the Frasnian these species are found in the high-latitude North American and Siberian Realms. Note the sharp latitudinal compression of the tethyan species to equatorial regions only in the Famennian. (From Kalvoda 1986, 1990. Artwork courtesy of J. Kalvoda.)

the Frasnian/Famennian boundary. Species of the high-latitude regions differentially survive the event, however (Kalvoda 1986, 1990). Most strikingly, species of the foraminiferid Tethyan Realm experience sharp contraction in their latitudinal distribution across the Frasnian/Famennian boundary (figure 6.4). They are found in high-latitude areas, far to the north and south of the equator, in the Frasnian. Species in the foraminiferid Siberian Realm occur as far north as 60°N latitude, using the paleogeographic reconstruction of Scotese and McKerrow (1990; see figure 5.11). Following the Frasnian/Famennian crisis, they are confined to low-latitude, equatorial regions of the Paleotethys Ocean only.

In the vagile benthos, the proetid, brachymetopid, and phacopod trilobites are present in Europe in the Famennian, whereas the North American representatives of those families virtually disappear in the Frasnian. Briggs and colleagues (1988) argue that this pattern of survivorship may be latitude related, with the selective survival of higher-latitude faunas. It is interesting that the survival of these trilobites in the eastern-facing shelf regions of Europe and their demise in

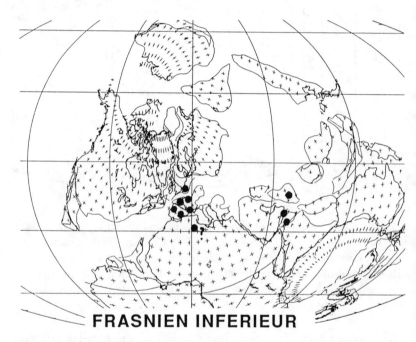

FRASNIEN INFERIEUR

Figure 6.5. Geographic distribution of asteropygine trilobites during the Frasnian, showing the absence of any high-latitude species within the group. (From Morzadec 1992; paleogeographic reconstruction used is that of Scotese and McKerrow 1990. See figure 5.11. Reprinted from *Evolution des Asteropyginae (Trilobita) et variations eustatiques au Dévonien* by P. Morzadec from *Lethaia* vol. 25, pp. 58–96, 1992, by permission of Scandinavian University Press. Artwork courtesy of P. Morzadec.)

the western-facing North American marine environments are similar to the geographic pattern of survival seen in the relict Frasnian stromatoporoid faunas (figure 6.2). Last, Morzadec (1992) argues for eustatic sea-level rise as the cause of the extinction of the asteropygine trilobites. Models of extinction triggered by sea-level rise will be considered in chapter 7. Here it is interesting simply to note the geographic distribution of the last asteropygine trilobites in the Frasnian (figure 6.5). They are all confined to the equatorial Paleotethys; they have no higher-latitude species members. And they do not survive the Frasnian/Famennian mass extinction.

BATHYMETRIC EFFECTS

In general, shallow-water marine ecosystems appear to be much more severely affected during the Frasnian/Famennian crisis than deeper-water faunas. Deep-water cnidarian, poriferan, and ammonoid species all show differential survival.

This bathymetric selectivity in extinction is seen most dramatically within the rugose corals, a group that suffers a massive loss in biomass at the Frasnian/Famennian boundary. Only 4% of the shallow-water species survive the biotic crisis. Deeper-water species suffer a 60% extinction in their numbers, and although this reduction is severe, it pales in comparison with the 96% loss of species in the shallow waters (figure 6.6). The decimation of the shallow-water corals is actually more severe than that of the stromatoporoids (Stearn 1987; Oliver and Pedder 1994).

A particularly intriguing bathymetric pattern of selective extinction and diversification occurs across the Frasnian/Famennian boundary in the Appalachian marine ecosystems of eastern North America, mentioned in the beginning of this chapter. Simultaneously with the extinction of many shallow-water benthic species the hexactinellids (glass sponges) migrate from deeper-water regions into the shallows and undergo a burst of diversification in species numbers (figure 6.7). Modern glass sponges are generally found in water depths in excess of 200 m and are considered to be better adapted to colder waters than most other invertebrate species. When the shallow-water species

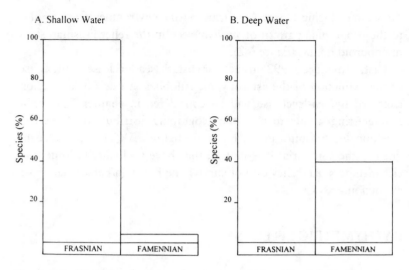

Figure 6.6. Diversity of species of rugosan cnidarians that inhabited marine shallow-water (A) and deep-water (B) habitats during the Frasnian and Famennian. (Data from Pedder 1982 and Sorauf and Pedder 1986.)

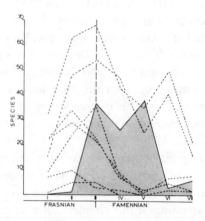

Figure 6.7. Reciprocal diversity changes in shallow-water marine invertebrate species (dashed lines) and deep-water hexactinellid sponges (solid line and shaded region) during the Frasnian and Famennian in New York State. The glass sponges actually blossom while other species decline, and decline when shallow marine species begin to recover in the late Famennian. (From McGhee 1982.)

begin to recover diversity in the late Famennian, the deep-water hex-actinellids show a reciprocal decline in diversity (figure 6.7).

Within the nekton, House (1988a) notes that the tornoceratid ammonoids selectively survive the Frasnian/Famennian crisis. The tornoceratids inhabited colder and deeper waters than many other ammonoid groups, and it is the tornoceratids that successfully diversi-fied following the demise of other ammonoids in the extinction event (House 1988a).

In common with other extinction events in the Earth's history, the upper oceanic water habitat of the marine plankton is massively disrupted (Tappan 1982). Approximately 90% of the preservable phytoplankton is affected during the Frasnian/Famennian crisis; massive biomass reductions also occur among the zooplankton. Something goes seriously wrong for organisms living in the surface marine waters of the Earth during the Frasnian.

MARINE VERSUS TERRESTRIAL HABITAT EFFECTS

We saw in chapter 4 that both the sea and land are affected in the Frasnian/Famennian mass extinction. To judge the severity of the extinction between these two major divisions of the Earth's surface is not simple, as they are ecologically so different from one another. Fortunately, we have a group of animals that inhabited both worlds in the Late Devonian: the fish.

A marked habitat effect in selective survival can be observed in the Devonian fish groups that have both marine and freshwater species members. The two main groups of interest are the acanthodians and the placoderms (figure 6.8). Curiously, they show a proportional response to the mass extinction that is very similar in magnitude, even though the placoderms are an order of magnitude more speciose than the acanthodians. Both groups lose a little more than half of their species.

Of importance, however, is the fact that in both groups it is the marine species that are differentially lost (figure 6.8). Only 12% of the acanthodian marine species survive, in contrast to a 70% survival of the freshwater species. The numbers are similar for the placoderms: 35% survival of the marine species versus 77% survival of the fresh-

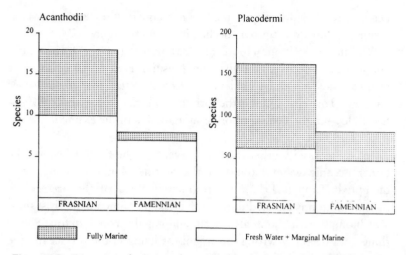

Figure 6.8. Diversity of species of acanthodian and placoderm fishes that inhabited fully marine habitats (shaded histogram portion) and freshwater to marginal marine habitats (unshaded histogram portion) during the Frasnian and Famennian. (Modified from McGhee 1982.)

water species. Furthermore, Long (1993) demonstrates that for three of the surviving families of placoderms that have species inhabiting both freshwater and marine habitats in the Frasnian (the Bothriolepididae, Asterolepididae, and Groenlandaspidae), it is only the freshwater species that survive into the Famennian. He estimates that fully one-third of the surviving placoderm fish families leave marine environments entirely as a result of the ecological selectivity of the Frasnian/Famennian mass extinction.

A key environmental parameter that differentiates marine from freshwater habitat regions (other than salinity, of course) is temperature. In general, terrestrial freshwater species are adapted to seasonal and diurnal fluctuations in temperature. Your average freshwater fish must tolerate much wider absolute ranges in temperature, and much-shorter-term fluctuations in temperature, than a marine fish. In contrast to terrestrial rivers and lakes, shallow marine regions are temperature buffered, largely because of the thermal inertia of the great oceanic water masses with which they are in contact. Marine fish experience much smaller changes in absolute temperature, and what temperature changes do occur do so very slowly. Thus the differential survival of terrestrial freshwater fish species may reflect their greater tolerance of temperature changes (McGhee 1982).

GEOGRAPHIC EFFECTS?

Widespread geographic distribution is often cited as a survival factor in mass extinction (Jablonski 1986a,b; 1989; 1991). Species whose populations are confined to a small geographic range are obviously at higher risk of extinction than species whose populations are found all over the planet. Highly endemic species, such as lemurs in Madagascar or koalas in Australia, could easily be lost as they exist in low population densities to begin with and are found nowhere else. A local environmental catastrophe, whether it be an asteroid falling from the sky or humans cutting down all the trees, will eliminate these species. Humans, on the other hand, are found in populations on every continent of the Earth (even frozen Antarctica, if one counts the various scientific research teams that work there). No merely local event could possibly eliminate the human species.

The reasoning is ecologically sound. Thus does a geographic signal also exist in the Frasnian/Famennian data outlined in chapter 4? Do species with widespread populations differentially survive and endemic species perish in the Frasnian/Famennian crisis?

The data appear to say no. The cnidarian genera of the Frasnian are cosmopolitan; that is, they have populations scattered in suitable habitats around the planet. This was not always the case. In the Early and Middle Devonian cnidarian generic populations are highly endemic, being divided into two great realms (the Old World Realm and the Eastern Americas Realm), which themselves are subdivided into smaller geographically distinct provinces (Oliver and Pedder 1979; Oliver 1990; see figure 5.2). This endemism breaks down with rising sea levels in the Frasnian, and by the end of the Frasnian the same genera, and some species, of cnidarians are found worldwide. To repeat: Frasnian cnidarian species have widespread geographic distributions indeed. And the cnidarians are decimated in the Frasnian/Famennian mass extinction.

The same bleak picture is seen for other organisms. Stromatoporoid faunas are highly provincial in the Early Devonian, but by the

Middle and early Late Devonian they are cosmopolitan in distribution (Stock 1990, 1993). Their widespread geographic distributions do not save them from the Frasnian/Famennian mass extinction. Brachiopods are highly endemic in the Early and Middle Devonian, subdivided into three geographic provinces (the Old World, Appalachian, and Malvinokaffric; Johnson and Boucot 1973; see figure 5.1). Brachiopod endemism also breaks down by the Late Devonian (Johnson and Boucot 1973; Boucot 1975, 1988). Extensive migration of brachiopod species occurs beginning in the early Frasnian, and many local endemic faunas are progressively replaced through the span of the Frasnian by cosmopolitan species immigrants (McGhee 1981b, 1995). Late Frasnian brachiopod genera have very widespread geographic ranges, but still are decimated in the Frasnian/Famennian mass extinction. Land plants are divided into six floral provinces in the Middle Devonian, but by the Frasnian plant endemism has declined so that only two, or possibly three, floral provinces are left (Scheckler 1984; Raymond 1993). Frasnian land plants have wide geographic distributions, but they still suffer the "diversity minimum" of the Frasnian/Famennian crisis (Raymond 1993).

Other factors seem to be more important in differential survival of the Frasnian/Famennian crisis than absolute extent of geographic range. Stromatoporoid and brachiopod genera with species distributed all through the wide geographic extent of the Late Devonian Paleotethys Ocean still suffer extinction. Only if some of those genera extend up into the higher latitudes, away from the equator, do we begin to see diversity losses that are not as severe.

TROPHIC EFFECTS?

Detritus feeding has also been argued to be a buffer to mass extinction, that is, to be in a position in the food chain that is not directly tied to living plant matter may be an advantage (Sheehan and Hansen 1986). The base of the trophic pyramid consists of the photoautotrophic producers, the phytoplankton and benthic plants in the oceans and terrestrial plants in rivers and lakes and on the land. If the plants die, the herbivores that eat those plants also die, and the carnivores that eat the herbivores die as well. Destruction of the base of the trophic pyramid collapses the upper ecological levels as well. Sheehan and Hansen (1986) and Van Valen (1984) argue that scavenging organisms differentially survive mass extinction. Organisms in the food chain directly dependent upon living plant matter die if photosynthesis ceases in an environmental catastrophe, whereas scavenging and detritus-feeding organisms may be able to survive on dead plant material (plus other dead organisms) until photosynthesis resumes with plant recovery. Sheehan and Hansen (1986) note, in support of this hypothesis, that benthic scavengers and detritus feeders in the oceans and freshwater ecosystems differentially survive the Cretaceous/Tertiary mass extinction. The marine filter-feeding organisms and the zooplankton, and the carnivores that eat those organisms, are all dependent on living matter, and they all differentially suffer extinction.

Rhodes and Thayer (1991) put a different spin on the trophic effects hypothesis by arguing that organisms that are able to tolerate outright starvation should better survive mass extinction. They note that filter-feeding brachiopods have minimal metabolisms compared to many filter-feeding molluscs, particularly those that also have higher-energy life-styles that include active locomotion in burrowing or crawling. They argue that the extinction rates of bivalves increase markedly, relative to brachiopods, in the Cretaceous/Tertiary mass extinction because of their higher energy (and hence food) requirements. Thus the filter-feeding brachiopods, which can tolerate long periods of starvation, did not suffer extinction proportionally with the filter feeding bivalves (Rhodes and Thayer 1991).

Do trophic effects play a role in the Frasnian/Famennian crisis? Do organisms that are scavengers and detritus feeders differentially survive the mass extinction, compared to organisms in food chains directly dependent upon living plant matter? Among the filter-feeding organisms, do organisms with minimal metabolisms survive relative to organisms with high-energy budgets?

The data appear to say no. Brachiopods with their minimal metabolisms are decimated in the mass extinction. Biconvex brachiopods that do survive the mass extinction tend to have more globe and equiconvex shellforms than preextinction brachiopods (McGhee 1996). Such shellforms have maximum internal volume to minimum shell surface area, which is the optimum geometry for organisms that feed using the lophophore system of filtration. The fact remains, however, that the high-energy mobile bivalves suffer much smaller diversity loss than the low-energy sessile brachiopods in the Frasnian/ Famennian mass extinction.

Among the bivalves themselves, those that are detritus feeders do not survive in greater numbers than those that are filter feeders (Rehfeld 1989). If anything, the reverse appears to be true, with detritus-feeding forms losing more species diversity than bivalve groups composed of filter feeders.

Food chains in freshwater ecosystems are often based on plant debris derived from the land. Thus one might expect freshwater trophic chains to be less disrupted in a mass extinction, as the ecologies

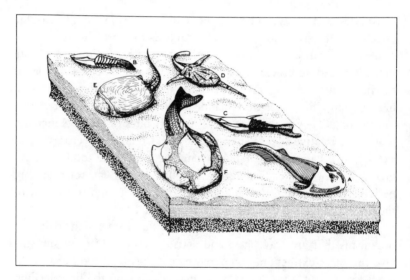

Figure 6.9. Devonian agnathan fishes: A = a cephalaspid, B = *Anglaspis*, C = *Pteraspis*, D = *Doryaspis*, E = *Zascinaspis*, F = an ostracoderm. Although detritus feeders, and common in freshwater ecosystems based on plant debris from the land, all fossil agnathans are exterminated in the Frasnian/Famennian mass extinction. Only the lampreys and hagfish survive today; thus presumably a few Frasnian agnathans must have survived somewhere (see chapter 4). (From Dineley 1990. Reproduced with permission of the Canadian Museum of Nature, Ottawa, Canada.)

of those organisms are detritus based. At first glance, this might appear to be a factor in the survival of freshwater fishes relative to marine fishes discussed earlier. On closer examination, however, this scenario does not hold up. The dominant detritus-feeding fishes in the Devonian are the agnathans (Dineley 1984, 1990). The agnathans suffer continuous diversity decline through much of the Devonian, probably because of increased predation from the evolving jawed fishes. Five subclasses of the agnathan fishes still exist in the Frasnian, and these interesting but bizarre-looking fish could be found around the world in freshwater streams (figure 6.9). Going fishing in the Frasnian world would have been an interesting experience, but one could catch these agnathans only in the Frasnian world. Detritus feeding did not benefit or save the Frasnian agnathans; they are entirely exterminated in the Frasnian/Famennian mass extinction (figure 4.17).

REPRODUCTIVE EFFECTS?

Finally, it has been suggested that reproductive mode may play a part in survival of mass extinction. Many of these scenarios have to do with differential survival of land vertebrates in the Cretaceous/Tertiary mass extinction, and variables of importance are brood (or litter) size, frequency of reproduction, degree of parental care, and so on.

The terrestrial ecosystem is just barely established in the Late Devonian. Thus many of the models for differential survival based upon reproductive effects do not apply. In the marine realm, however, it has been postulated that organisms with a nonplanktotrophic larval stage should differentially survive over those organisms having larvae that actively feed in the plankton. The plankton is hard hit in the Frasnian/Famennian crisis, and thus organisms with planktotrophic larvae should be decimated. (Even if the adult organisms themselves are unaffected, their young, and hence the next generation, are gone.)

Differential survival based upon reproductive mode does not appear to play a role in the Frasnian/Famennian crisis, as best as can yet be determined. The dominant shellfish of the Frasnian, the Brachiopoda, have lecithotrophic larvae, and many species actually brood their larvae. (This can be demonstrated by the presence of brood pouches in fossil brachiopod shells.) They should have differentially survived over many mollusc species with planktotrophic larvae. Instead the brachiopods are decimated.

SUMMARY: THE FIVE SIGNALS

Analysis of the record of taxa that witnessed the crisis, both victims and survivors of the Frasnian/Famennian mass extinction, in the context of their own world and ecology, reveals five nonrandom ecological signals in the data:

> *Signal 1:* The biosphere of the entire planet is affected, both the sea and the land.
>
> *Signal 2:* Life in equatorial and low-latitude regions of the Earth is decimated; diversity loss is not as severe in higher-latitude regions of the planet.
>
> *Signal 3:* Latitudinal compression of geographic range occurs in surviving low-latitude faunas. Faunas that previously extended their ranges up from the equator into midlatitude regions in the Frasnian lose these extensions of range in the diversity crisis and are restricted to equatorial regions in the Famennian.
>
> *Signal 4:* Life in shallow-water marine habitats is decimated; diversity loss is not as severe in deeper-water habitats.
>
> *Signal 5:* Clades having species that inhabit both marine and freshwater habitats (e.g., fish) experience severe diversity losses in their marine species; diversity loss is not as severe in terrestrial freshwater species.

In contrast to other mass extinctions, absolute geographic range appears not to be a factor in survival of the Frasnian/Famennian mass extinction. Nondependency on living plant matter also appears to confer no advantage or increased survival, nor does larval brooding in marine invertebrates. The significance and meaning of the five ecological signals that do emerge from the data will be discussed in chapter 7, where we finally confront the mystery of possible causes of the Frasnian/Famennian crisis.

VII

Proposed Causes of the Mass Extinction

W E NOW have arrived at the most difficult question to be asked concerning the Frasnian/Famennian mass extinction: What caused it? To know, to understand, the cause of a catastrophe removes some of its horror, in human affairs as well as in the great unfolding of the evolution of life on the planet. It is the mysterious catastrophe, the unexplained disaster, that disturbs us most. We can take no clear warning from the unexplained, make no preparation to defend the biosphere from the mysterious.

As paleobiologists, it would be professionally satisfying to be able to say to the world: Watch out! That is, we want to learn from our studies of the evolution of life in the past, the great natural experiment preserved in the fossil record, and to be able to make specific predictions about possible events in the future. The geophysicist would like to warn of impending earthquake, the meteorologist to warn of an approaching hurricane. Can we warn of impending mass extinction? Not without knowing what caused the ones that have already occurred in the Earth's history.

What can cause the massive loss in the Earth's diversity of life outlined in chapters 2 and 4? What lethal mechanism or mechanisms can operate on the timescales discussed in chapters 1 and 3? What environmental trigger could produce the ecological signature seen in chapter 6?

THERMAL: TOO COLD?

A proximate cause for the Frasnian/Famennian crisis, which has been argued to be most compatible with the observed ecological patterns of selectivity of the event (McGhee 1989b), is lethal temperature decline. Significant global cooling would decimate reefal and tropical ecosystems, as species inhabiting these regions have no refuge against cold. Higher-latitude ecosystems would be less affected, as many species could simply migrate to lower latitudes and thus maintain their tolerable temperature ranges.

Global cooling would most adversely affect the warm-water, temperature-buffered, epicontinental sea benthos. Deeper-water fauna are generally adapted to cooler water conditions, and thus would be expected to survive differentially. Shallow oceanic planktic ecosystems in equatorial regions would also be decimated, with greater survival of the plankton in high-latitude regions. Moreover, terrestrial ecosystems are adapted to, and thus more tolerant of, temperature fluctuations than marine. Each of these expected patterns of ecological selectivity in survival is seen in the Frasnian/Famennian crisis (signals 1 to 5, see chapter 6); thus lethal temperature decline is a prime candidate as the cause of the mass extinction.

Global Cooling: Glaciation?

A problem arises in determining the ultimate cause of global temperature decline, because the same temperature effect may be produced by a variety of forcing mechanisms, both terrestrial and extraterrestrial. The most obvious correlate with global cooling is continental glaciation; yet even here it can be debated whether the onset of glaciation has an extraterrestrial trigger (cyclic eccentricities in the orbit of the Earth around the sun) or a terrestrial trigger (plate tectonic positioning of a continental landmass over a pole), or both.

Caputo and Crowell (1985) and Caputo (1985) argue that the Frasnian/Famennian extinctions were triggered by the onset of glaciation in Gondwana. Their argument is based on two lines of reasoning: (1) reinterpretations of polar wandering paths for Gondwana, which have the south pole positioned on continental South America rather than over the ocean in the Late Devonian (see the more recent paleogeographic reconstructions in figure 5.11 and 5.12, which still have the pole on continental landmasses), and (2) new biostratigraphic miospore dates for glacial sediments in South America (see figure 5.8), which move these previously Carboniferous-dated sediments back in time to the early Famennian.

Several problems arise with the proposed Frasnian/Famennian glaciations. First, significant glaciation is usually correlated with marine regression due to ice volume buildup, but the late Frasnian and early Famennian are periods of maximum global sea level high stand (Johnson et al. 1985, and discussed in more detail later). Only late in the Famennian does significant marine regression occur.

Second, the early Famennian dates for South American glacial sediments are disputed by Streel (1986), who argues that the reported early Famennian miospores have been reworked upward into younger sediments of the latest Famennian age. More recent biostratigraphic work (Loboziak et al., in Streel 1992) with the South American glacial sediments confirms Streel's earlier charge and reveals that the sediments fall in the Middle *praesulcata* Zone of the standard conodont zonation (see table 1.6). As such, the South American glacial sediments are now dated in the very latest Famennian. Late Famennian to early Carboniferous glaciation would be consistent with the observed pattern of marine regression during this interval of time (McGhee and Bayer 1985; Johnson et al. 1985; Veevers and Powell 1987).

Last, new paleomagnetic data suggest that the south pole remains positioned over continental Gondwana for the entire interval of time from the Middle Ordovician to the Late Devonian (Van der Voo 1988; Miller and Kent 1988; Scotese and McKerrow 1990). Such a positioning brings into question the tectonic-glacial model of Caputo and Crowell (1985) and Caputo (1985), in which glaciation is initiated whenever a continent moves over a pole. Ormiston and Klapper (1992) maintain that there is no perennial snow cover even in the high latitudes of Gondwana during the Late Devonian, in spite of projected winter land temperatures as low as $-40°C$. This is due (in their model) to the lack of sufficient polar transport of moisture to form extensive snow cover, because there is less midlatitude cyclone activity during the Late Devonian than in the modern world.

In summary, the glaciers arrive too late. By the time of significant South American glaciation (Middle *praesulcata* Zone) the Frasnian/ Famennian crisis is history, as it occurred some 9 Ma previously. The late Famennian glaciation does appear interesting in that it may provide a possible trigger for the ammonoid diversity losses that occur in the Hangenberg event (see table 3.1).

Global Cooling: Paleogeography?

An intriguing alternative tectonic model of global cooling has been proposed by Copper (1986). In this model, Euramerica and Gondwana are separated by an ocean (the "Frasnian ocean"; see figure 7.1) in the Frasnian, an ocean that closes at the Frasnian/Famennian boundary. The closing of the "Frasnian ocean" disrupts the low-latitude circumequatorial flow of warm water that had previously existed in the ocean between these two supercontinents. Cessation of easterly equatorial currents would bring high-latitude cold water into equatorial regions along the western margins of this continental configuration, and would lead to restricted circulation with euxinic conditions in the surviving warm-water basins along the eastern margins (figure 7.1). The resultant cooler Famennian oceans and global climates are proposed to have eventually initiated glaciation on Gondwana later in the Famennian/Carboniferous transition.

This paleogeographic model of climatic cooling was, however, questioned in terms of both Late Devonian paleomagnetic data (Hurley and Van der Voo 1987; Van der Voo 1988) and biogeographic data (Raymond et al. 1987), both of which were stated to not support a Frasnian/Famennian collision of Euramerica and Gondwana. More recent paleomagnetic data do support such a collision, but maintain it occurs in the Early Devonian (Miller and Kent 1988; Scotese and McKerrow 1990; Dalziel et al. 1994; see the paleogeographic recon-

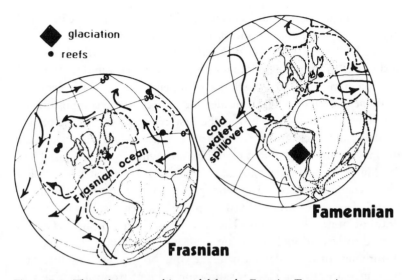

Figure 7.1. The paleogeographic model for the Frasnian/Famennian mass extinction of Copper (1986). In the Frasnian world, a large ocean is hypothesized to exist between Euramerica and Gondwana, with through- flowing warm-water equatorial currents. In the Famennian world, the ocean is hypothesized to have closed, resulting in cold water upwelling on the western margin of the supercontinent. The only surviving warm water reef faunas occur in the equatorial Paleotethys embayment, and glaciers are postulated to occur in South America. (From Copper 1986. Used with permission of the author.)

structions of figure 5.11 and 5.12). Either way, the timing of the paleogeographic model for global cooling, because of disruption of circumequatorial water flow, does not work for the Frasnian/Famennian crisis interval. Similarly, anoxic water conditions existed sporadically throughout the entire Frasnian (and not just in the eastern equatorial geographic region shown in figure 7.1), and not just at the Frasnian/Famennian boundary.

Global Cooling: A Shortage of Atmospheric CO_2?

A major concern in the modern world is a surplus of atmospheric CO_2, caused by human burning of fossil hydrocarbons as fuels, and its possible global warming effects. Could the opposite have occurred in the Late Devonian, in a sort of reverse greenhouse effect?

The Devonian is the *Age of Land Plants.* The invasion of the land surfaces of the Earth by photoautotrophs, and their rapid (in terms of geologic time) spread across the continents during the Devonian, could be expected to have considerable impact on the composition of the planet's atmosphere, just as the first evolution of aerobic photosynthesis did back in the Precambrian. Algeo (1994) maintains that the massive increase of land plant biomass, which occurs in just 15 to 20 Ma in the Middle and Late Devonian, precipitates a "global biogeochemical crisis." The land plant biomass increase results in burial of organic carbon (fixed by plants), which results in a downdraw of atmospheric CO_2. The downdraw in atmospheric CO_2 is argued to lead to global cooling (a reverse greenhouse effect) and a disturbance of the oceanic carbonate equilibrium, resulting in short-term increases in the $CaCO_3$ saturation state of tropical marine waters in particular. Growth of the massive reefs that exist in the Frasnian is seen as a result of this saturation state, which is argued to precipitate in turn a massive depletion of seawater bicarbonate by the time of the Frasnian/Famennian boundary. That depletion is argued to be reflected in the extinction of calcifying reef organisms at this time because of a "carbonate crisis" (see Narkiewicz and Hoffman 1989). Thus the peak of the "global biogeochemical crisis" is seen as the Frasnian/Famennian mass extinction (Algeo 1994). This model was further refined by Algeo and Maynard (1994) to include oceanic anoxia, by arguing that Frasnian shallow marine anoxic phases were driven by episodic nutrient pulses from the land plant biomass, producing eutrophic conditions in shallow marine areas.

Robinson (1990) also notes the massive effect the evolution of land vegetation must have had on the Earth's atmosphere at this time, in terms of both decline in CO_2 partial pressures and increase in atmospheric O_2 levels. She believes these changes, though major, are distributed over the entire Devonian in a progressive fashion, building to a peak in the Late Carboniferous—that is, a progressive decline in atmospheric CO_2, and increase in O_2, spread over some 100 Ma of time.

Berner (1990) and Mora and colleagues (1991) agree that partial pressures of CO_2 in the atmosphere are much higher in the Early to Middle Paleozoic than at present, on the basis of both atmospheric modeling (Berner 1990) and empirical measurements of carbon isotope compositions of Silurian and Devonian sediments (Mora et al. 1991). However, Berner (1990) also shows the decline of atmospheric CO_2 to be a long-term event, from a maximum in the Cambrian to a minimum in the Late Carboniferous. The steepest slope of the decline function does occur in the Devonian (figure 7.2). Specifically, a sharp falloff in CO_2 values occurs in the interval of time from the Middle Devonian into the Early Carboniferous, or some 40 Ma. The Frasnian/Famennian boundary occurs at about the midpoint of this segment of the decline function (figure 7.2).

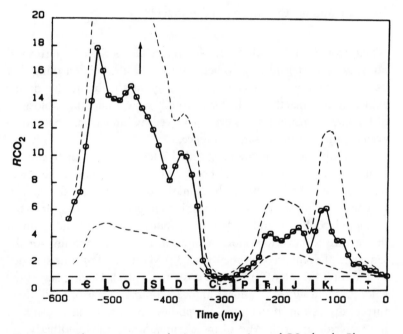

Figure 7.2. Changes in atmospheric concentrations of CO_2 for the Phanerozoic, according to Berner (1990). CO_2 levels are very high in the Cambrian, then decline to minimum values in the Carboniferous. Note that the Late Devonian falls on the steepest segment of the decline function. Symbol key: RCO_2 = ratio of CO_2 in the ancient atmosphere relative to that in the modern, solid line with circles = best estimate value for ancient RCO_2, dashed lines = envelope of approximate error of estimation. (From Berner 1990. Copyright (c) 1990 by the American Association for the Advancement of Science. Used with permission.)

Clearly, the evolution of land plants in the Devonian has a massive effect on the further evolution of the planet's atmosphere. That effect appears to have been a long-term one, but cumulative and spread over millions of years. The plant-driven atmospheric changes simply do not appear to be abrupt enough, and intense enough, to account for the Frasnian/Famennian mass extinction. The very cumulative and continuous nature of the atmospheric changes cannot be reconciled with the several discrete and temporally separated extinction pulses that take place in the 3.0 Ma crisis interval surrounding the Frasnian/Famennian boundary (table 3.2 and figure 3.5).

Global Cooling: Extraterrestrial Impact?

When people think of the possible impact of a large extraterrestrial body, say an asteroid or a comet, with the Earth, the first thing that usually comes to mind is the immensity of the blast, the titanic explosion of impact. And the blast would be awesome; the explosions of the largest fusion bombs ever produced by humans are not worth even mentioning in the same sentence.

Curiously, however, the most lethal effect of a hypothetical collision of a large extraterrestrial body with the Earth is not the direct result of the blast, but rather the climatic consequences of the impact. One way to chill the planet very fast is to impact an asteroid or two.

Global cooling, produced by large-body impact, would seem at first glance to have too short a duration in time to account for the sequence of events seen in the 2.0 to 3.0 Ma interval of time spanned by the total Frasnian/Famennian crisis (table 3.2). Having stated the chief problem with the impact-induced cooling scenario, I will defer further discussion of the impact hypothesis here and will return to it in greater detail at the end of the chapter, where the hypothesis will be examined in its own individual section.

THERMAL: TOO HOT?

Temperature is a pervasive ecological factor. Organisms that may be able physiologically to tolerate the direct effects of temperature change may succumb to the indirect effects, changes in gas solubility in marine waters, changes in nutrient flow, changes in oceanic currents, and so on. In a review of the effects of temperature in marine organisms, Clarke (1993) outlines the complexity of ecological factors dependent upon temperature and remarks: "I suspect that few paleoecologists would ascribe temperature-associated extinction to simple freezing (or cooking) to death" (Clarke 1993:511). Yet in the same paper he outlines a local extinction event in which the species concerned was literally cooked to death. In the 1982–1983 El Niño anomalous warming event, the dominant bivalve *Mesoderma donacium* was entirely eradicated along the Pacific coast of South America. This species had previously been present in densities of 3000 individuals/m^2 and comprised 90% of the bivalve fauna (Clarke 1993). In 2 short months in 1983 all the individuals of the species died. The hot waters of El Niño had been too hot for *Mesoderma donacium*.

Thompson and Newton (1988) propose the Frasnian/Famennian mass extinction to be driven by lethal temperature increase, and not decline. They note that many marine organisms live closer to the upper temperature limits of their physiologic tolerances than to their lower. Thus smaller changes may be required to bring about mass death in the increased temperature direction than in the decreased direction.

The Late Devonian marine realm is characterized by high sea level stands, widespread anoxic bottom waters, and warm surface temperatures. Thompson and Newton (1988) propose to drive these temperatures even higher, into the upper lethal limits. Because oxygen would be less soluble in the hotter marine waters, the oxygen minimum zone would expand, or, as Thompson and Newton (1988) put it, marine organisms would be lethally trapped between anoxic waters from below and hot surface temperatures from above.

Support for the high-temperature scenario comes from carbon and oxygen isotopic studies of brachiopod shell material conducted by Brand (1989). Ratios of the oxygen isotopes ^{18}O:^{16}O are temperature sensitive, and Brand (1988) found oxygen isotope ratios preserved in Devonian brachiopod shells that indicate an increase of temperature during the Late Devonian to over 60°C at the Frasnian/Famennian boundary. From the isotope data, the entire Devonian would have temperatures with mean ranges from 36°C to 54°C.

Something is clearly wrong with these data, as Brand (1989) notes. The average lethal thermal limit for most marine invertebrates is around 38°C; thus if the preceding values were true, nothing should have been alive in the ocean for just about the entire Devonian. Yet life clearly flourishes in the Devonian. Thus Brand (1989:321) introduces an adjustment to bring the values into "*realistic* environmental and biological conditions" (italics Brand's). Following this "adjustment," temperature ranges fall into the livable range except at the Frasnian/Famennian boundary, where they climb into the lethal range of 40°C (Brand 1988).

Needless to say, these adjusted temperature data have been met with skepticism. Oxygen isotopes are also notoriously susceptible to diagenetic alteration, as will be discussed in chapters 8 and 9. Other published isotopic studies of brachiopod shell material show wide ranges of scatter in the data for the entire Devonian (Popp et al. 1986; Wadleigh and Veizier 1992). Further isotope stratigraphic work needs to be done to determine the true $^{18}O:^{16}O$ composition standards for the ancient Paleozoic oceans, work that has been termed "a precondition for a more rational discussion of oxygen isotopic properties of past seawater, a topic of high emotional content in the extant research community" (Wadleigh and Veizier 1992:442).

Short-term El Niño-style warming events are clearly lethal but also clearly local. In order to trigger the magnitude of lethality seen in the Frasnian/Famennian mass extinction, the entire planet must become hotter. Thompson and Newton (1988) mention the possibility of CO_2-induced global greenhouse warming, but the Late Devonian atmosphere contains less CO_2 than the earlier Devonian, and CO_2 levels are declining rapidly during this interval of geologic time (Berner 1990; see figure 7.2). Further, the Frasnian/Famennian crisis was not a single die-off event, but a series of events. To circumvent this difficulty, Becker and House (1994) propose a series of "periodic 'greenhouse' climatic overheating episodes" around the Frasnian/Famennian boundary. The putative global triggering mechanism for these episodes is not specified, perhaps wisely, as such episodes could not be due to the conventional greenhouse mechanism of atmospheric CO_2 increase.

Last, and most important, the ecological expectations of the temperature increase model do not match the observed ecological signature of the Frasnian/Famennian mass extinction. In a much hotter world, reefal ecosystems should have flourished and all tropical species should have expanded their ranges into high latitudes. The exact opposite is seen in the Frasnian/Famennian mass extinction (signals 2 and 3, chapter 6).

CLIMATE: TOO SEASONAL OR TOO UNSEASONAL?

In an early study Frakes (1979) maintained that the entire Late Devonian world was warmer and dryer than at the present, which produced widespread evaporitic conditions across the planet. This reconstruction was based in part on the known distribution and concentration of halite and other evaporitic sedimentary deposits in the geologic record, which are indeed greater in the Devonian than in either the Silurian or Carboniferous (Kalinko 1974; Gordon 1975).

This earlier climatic reconstruction has been called in question, as it is now known that global temperatures are cold enough in the latest Famennian (the Middle *praesulcata* Zone) to initiate continental glaciation in the South American region of Gondwana (Streel 1986, 1992). The geographic distribution of late Famennian glacial striations, tillites, and other glacial sediments in the Amazonas Basin is extensive, indicating that the glaciation is not a mere local event (Caputo and Crowell 1985; Caputo 1985). Other signals of the onset of cold conditions in the late Famennian are the sharp sea level falls, presumably driven by ice volume buildup on Gondwana, which occur during this interval of time (see figures 7.3 and 7.9).

In a more recent summary of global paleoenvironments, Frakes and colleagues (1992) postulate that global temperatures rise through the Silurian into the Devonian, and then fall beginning in the middle of the Carboniferous, with the onset of glacial conditions occurring in the Late Carboniferous (figure 7.3). The late Famennian glaciation episode (see the preceding discussion) is now shown as a brief dip in the overall high values of the mean global temperature curve. Although geographically extensive, the late Famennian glacial episode is considered by Frakes and colleagues (1992) to be of short duration, and they continue to characterize the Late Devonian world as one considerably and consistently warmer and dryer than the present.

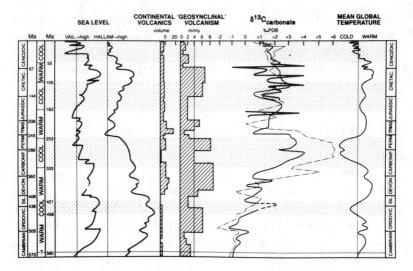

Figure 7.3. Mean global temperatures (curve at far right) of the Earth through the span of the Phanerozoic, according to Frakes et al. (1992). The entire Devonian is shown as hotter (temperature curve to the right of the center line, which represents present global mean temperature) than the present world, with evaporitic conditions present during the Late Devonian in particular (Frakes 1979:99). (Figure copyright (c) 1992 by Cambridge University Press. Reprinted with the permission of Cambridge University Press.)

What was the climate like during the Late Devonian? Tectonic models of climatic conditions during the Paleozoic suggest a Late Devonian world with a much more unstable and changeable climate than thought previously.

Climate: Tectonic Megacycles?

Fischer and Arthur (1977) proposed a climatic model for the entire Phanerozic based upon changes in the internal dynamics of the Earth's mantle and core, and the effect of those changes on the surficial climate of the planet. They divide the past 600 Ma into five major climatic maxima and minima in which the Earth moves from a cold "icehouse" condition to a warm "greenhouse" condition before starting the cycle all over again (figure 7.4). Greenhouse conditions

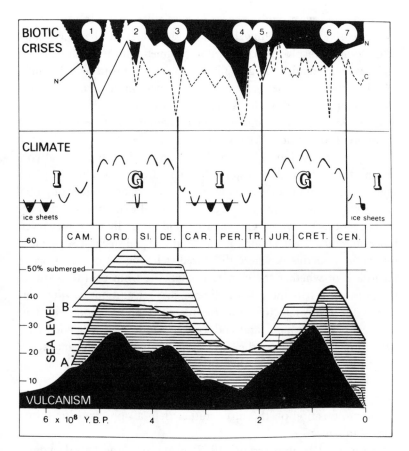

Figure 7.4. Megacycles in global climate for the entire Phanerozoic, proposed by Fischer and Arthur (1977). The Earth's climate is shown as oscillating between "I" ("icehouse") and "G" ("greenhouse") states in the sinusoidal curve in the center of the figure, with which biotic crises (top part of figure), volcanic activity (bottom part of figure), and sea level changes (lines A and B at bottom) are correlated. The Late Devonian falls on the steep slope of the sinusoidal climatic curve, in the transition from the greenhouse world of the Silurian to the icehouse world of the Carboniferous. (From Fischer 1984. Used with permission of the author.)

are associated with major phases of accelerated plate-tectonic activity: extensive magma upwelling and spreading at the midoceanic ridges (which displaces ocean water volumes onto the continents in high sea level stands) and globally increased volcanism (which pumps CO_2 into the atmosphere). The reverse happens in the icehouse phase: Plate-tectonic activity slows down, the oceans drain off of the continents in low sea level stands, and the atmosphere becomes more transparent to heat loss because of lower CO_2 levels.

Fischer and Arthur (1977) make a very intriguing proposal concerning the state of the biosphere of the Earth in their model. They suggest that rapid climatic change itself is the major trigger of global ecosystem collapse, and not the environmental extremes of either the cold icehouse or warm greenhouse conditions. In their megacycle function, climate is changing most rapidly at the steepest segment of the curve, and this is where they predict biological crises to occur, regardless of whether the climate is changing from cold to warm or warm to cold. It is the rate of climate change, and not its direction, that is the critical factor. And the match is pretty good. The Late Devonian, end-Permian and end-Triassic, and end-Cretaceous mass extinctions all occur on the steep-slope segments of the megacycle curve (figure 7.4). Smaller diversity crises in the Late Cambrian and Cenozoic also occur in this curve position. Of the "Big Five" mass extinctions, only the end-Ordovician is anomalous and falls at a maximum position on the curve, rather than a slope position (biotic crisis 2 in figure 7.4).

In the Fischer and Arthur (1977) megacycles the Late Devonian falls on the steep slope of the climatic curve in the transition from a warm hothouse world to a cold icehouse one (biotic crisis 3 in figure 7.4). The shape of their climate curve for this interval of time is very similar to the atmospheric CO_2 curve of Berner (1990), where the Late Devonian also falls on the steep slope of the CO_2 descent from high partial pressures in the earlier Devonian to low partial pressures in the Carboniferous (figure 7.2). Additional support for the model comes from the apparent frequency of intense storms through geologic time (Brandt and Elias 1989). These models would predict a world in which the global climate changes dramatically from the Frasnian through the Famennian.

Climate: Pulsation Tectonics?

A much more rapidly fluctuating tectonic model of climate has been proposed by Sheridan (1987). In Sheridan's model the Earth oscillates between periods of activity and quiescence at the mantle/core boundary deep within the Earth. In active phases, hot plumes rise from the mantle/core boundary and drive accelerated plate-tectonic activity at the surface of the Earth. The core becomes cooler and convects smoothly, producing a long period of stable polarity in the magnetic field of the Earth. In quiescent phases, no plume eruptions occur and the core becomes hotter (as it is shedding little heat) and convects more turbulently, producing numerous reversals in the magnetic field surrounding the planet. Sheridan (1987) proposed that six maxima and minima in superplume activity occur in the Paleozoic alone (figure 7.5). Other predicted phases of superplume activity in the Mesozoic and Cenozoic have been confirmed by Larson (1991).

The Paleozoic megacycle of Fischer and Arthur is replaced by three superplume cycles in Sheridan (1987). The Late Devonian falls in a

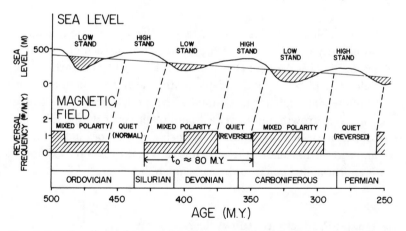

Figure 7.5. Smaller scale climatic cycles proposed for the Paleozoic by Sheridan (1987). The Late Devonian falls during a magnetic quiet zone with reversed field polarity and high sea levels, associated with active plume eruptions at the Earth's mantle/core boundary. (From Sheridan 1987. Artwork courtesy of R. E. Sheridan.)

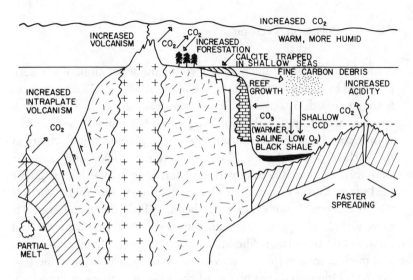

Figure 7.6. Planetary surface climate conditions during an active plume erup-
tion phase in the Earth's interior, according to the model of Sheridan (1987).
Late Devonian climates are predicted to be warm and humid, with high organic productivity in both the sea and on the land. (From Sheridan 1987. Artwork courtesy of R. E. Sheridan.)

zone of stable (but reversed to the modern field orientation) magnetic
field polarity (figure 7.5). That is, in the pulsation-tectonics model of
Sheridan (1987) the Late Devonian world should be characterized by
accelerated plate-tectonic activity, with warm and humid climatic
conditions and high sea level stands. The model predicts, in general,
greater reef growth and increased land plant cover during this phase
of superplume activity (figure 7.6). Such a world does indeed appear
similar to the Frasnian, at least as it is seen in the fossil record. It does
not appear at all similar to the Famennian, however.

Climate: Productivity Autocycles?

Buggisch (1991) expands upon these tectonic models and presents an
"autocyclic" model for Late Devonian climates, which seeks to factor
in the additional effects of biotic productivity changes and organic

carbon distribution (figure 7.7). In this model, organic productivity is postulated to increase greatly in the warm, shallow epicontinental seas associated with marine transgressive episodes. Simultaneously, warm surficial waters tend to promote oceanic stagnation at depth, as dense water no longer forms in the polar regions if temperatures at high latitudes do not fall below 5°C (Wilde and Berry 1986). The formation of dense water at the poles then flows in deep-sea currents toward the equator, and this ventilates the deep ocean. The combination of high organic productivity at the surface and stagnant water at depth is seen to produce vast regions of anoxic water masses, with major deposition of organic carbon in marine sediments.

The climatic effect of high productivity is postulated to be global cooling, because of the removal of a significant fraction of the CO_2 content of the atmosphere, which tends to warm the planet via the greenhouse effect. The initiation of an icehouse phase of the cycle is seen to trigger continental glaciation, and the resulting buildup of ice at high latitudes to trigger marine regression in turn (figure 7.7). Ocean basins become ventilated once again, with the reestablishment of deep-sea currents driven by the flow of dense polar sea waters. Organic carbon is reintroduced into the atmosphere and oceans via weathering of sediments exposed by falling sea levels and by oxidation of reduced carbon in unlithified marine sediments. Buildup of carbon dioxide levels in the atmosphere is postulated to initiate the next greenhouse phase of the cycle (figure 7.7).

Buggisch (1991) suggests that such an autocyclic climatic process would produce a characteristic sedimentary and geochemical signal in the geologic record. Following the transgressive phase, sediments rich in organic carbon should be deposited and positive shifts in $\delta^{13}C$ should be seen in pelagic carbonates ($\delta^{13}C$ is a measure of the ratio proportion of the stable isotopes of carbon and is discussed in more detail in chapter 8). The positive shift in $\delta^{13}C$ is driven by large amounts of the lighter isotope of carbon, which is preferentially taken up by organisms, being deposited in organic-rich anoxic sediments and therefore removed from the system. This phase should in turn be followed by marine regression and a negative shift in $\delta^{13}C$, indicative of the reintroduction of organic carbon by surficial weathering and marine oxidation.

In this model Buggisch (1991) maintains that climate in the Late

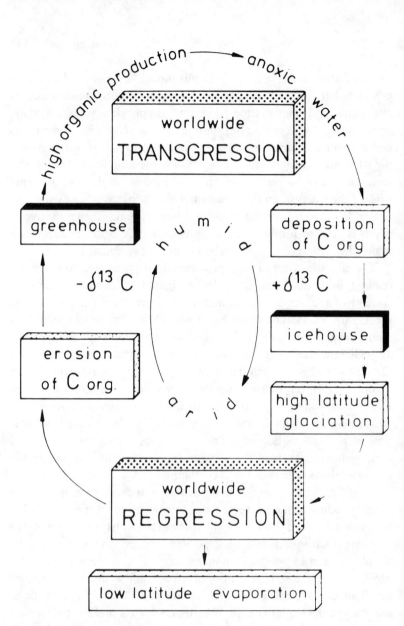

Figure 7.7. The "autocyclic" climatic model for Late Devonian climates, proposed by Buggisch (1991). Model climates oscillate between two extrema, one extremum with hot and humid conditions and associated marine transgressions and anoxic bottom waters, the other extremum with cold and dry conditions and associated marine regression and high-latitude glaciations. Deposition or erosion of organic carbon ("C org") in the cycle is hypothesized to produce positive and negative fluctuations in $\delta^{13}C$ in the world's oceans. (From Buggisch 1991. Used with permission of the author.)

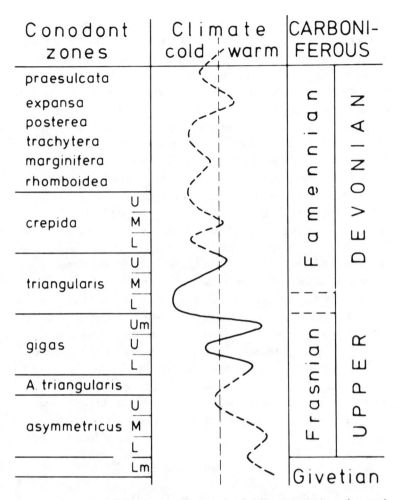

Figure 7.8. Proposed climatic oscillations for the Late Devonian, referenced against conodont zonal position (see tables 1.6 and 1.7). Eight cycles are proposed, the coldest and longest of which is proposed to occur at the Frasnian/Famennian boundary (see table 1.7 for conodont zonal comparisons). (From Buggisch 1991. Used with permission of the author.)

Devonian oscillates frequently between warm and cold phases (figure 7.8) and, rather than being extendedly hot and dry (figure 7.3) or hot and humid (figure 7.6), is in fact more similar to the variable climates of the late Cenozoic.

Buggisch (1991) attributes the Frasnian/Famennian mass extinction to the two climatic autocycles that occur near the boundary between the two stages (figure 7.8). In a sort of a one–two punch, organisms are first hit with hot climates, marine transgression, and spread of anoxic bottom waters across the shallow marine regions of the world. Those organisms that survive the first punch are hit with another, as climates rapidly cool and seawaters drain away from the continents in a major marine regression. Specifically, he maintains this one–two climatic couplet happens twice, first in the Late *rhenana* Zone, producing the peculiar Lower Kellwasser Limestone in Europe, and again in the *linguiformis* Zone, producing the Upper Kellwasser Limestone. In the span of 1.5–2.0 Ma organisms are hit with two cycles of heat/oxygen deprivation (in transgression) followed by cold/ habitat loss (in regression). Whatever hardy organisms survive the first hot/cold autocycle in the Late *rhenana* Zone are finished off by the second hot/cold autocycle in the *linguiformis* Zone, leaving devastation in the early Famennian.

The autocyclic model predicts many such environmental oscilla- tions throughout the Late Devonian; thus it is not at all clear why the two near the Frasnian/Famennian boundary should have such a devastating effect. Buggisch (1991) draws these two with larger cycle amplitudes (figure 7.8), but does not explain why this is so. More- over, the model is *autocyclic*; that is, it drives itself with cycles of organic productivity. If the photosynthesizers are almost wiped out, as they are in the Frasnian/Famennian mass extinction, then the entire cycling mechanism (figure 7.8) should come to a halt in the Fa- mennian.

Climatic Models: A Summary for the Late Devonian

As can be seen above, there exists little agreement concerning Late Devonian global climatic systems. Attributing the Frasnian/Famen- nian mass extinction to the evolution of the intrinsic climatic system of the Earth appears difficult. The shorter-term climatic cycles in the Buggisch (1991) and Sheridan (1987) models cannot drive mass extinction; otherwise a mass extinction would occur every time a climatic oscillation is hypothesized to occur, and it does not.

Perhaps the best candidate for a climate-driven mass extinction scenario remains the original Fischer and Arthur (1977) model. The positioning of the Late Devonian on the steep slope, or most rapidly changing, segment of their climatic curve matches the independent predictions of similar rapid changes in atmospheric CO_2 content made by Berner (1990) for this same interval of time. Crowley and North (1988) have modeled what they consider to be potential conditions for abrupt, discontinuous, and catastrophic changes in the Earth's climate. They believe that changing boundary conditions, such as the major shift from a greenhouse world to an icehouse world, can "cause discontinuous responses in climate models and result in relative rapid transitions between different climate states. Such terrestrially induced abrupt climate transitions could have contributed to biotic crises in earth history" (Crowley and North 1988:996). The key word in all of the preceding discussion is *models*. Theoretically, it may be possible to drive mass extinction with intrinsic changes in the Earth's climate—but it is difficult to prove it.

MARINE REGRESSION: VANISHING HABITATS?

Sea level fall has long been argued to correlate with times of biotic crisis in the geologic record (Hallam 1981; Flessa et al. 1986). As marine regression is also produced with ice volume buildup, separating the environmental effects of temperature decline from sea level fall can be difficult. In geologic periods where marine regression occurs without demonstrable glaciation, the causal link between lowered sea levels and elevated extinction levels is still problematic (Jablonski 1986a; Stanley 1987; McGhee 1989a). Further, many marine regressions occur in Earth history with no obvious biological effect seen in the fossil record. This last observation may be expected, however, as is discussed later.

An often-cited causal link between marine regression and extinction is the species–area effect. Marine regression clearly reduces the amount of habitable area on the continental shelves for shallow marine species. A reduction in habitat space would thus be expected to produce decreased population sizes in the marine benthos and to result in the eventual extinction of many of these species. The most serious crisis in the history of life on the Earth, the end-Permian event (figure 1.2), has been argued to be solely due to the habitat reduction effect of the massive marine regressions and low sea level stands that characterize the planet during that interval of time (Schopf 1974; Simberloff 1974). Major marine regression also occurs during the end-Ordovician event (figure 1.2), but that sea level fall was triggered by continental glaciation, thus compounding the effects of habitat destruction and those of global refrigeration (Brenchly 1989). There are several problems with the species–area scenario, however. First, although marine regression produces habitat reduction in the marine realm, the areal extent and habitat space of the terrestrial realm increases. Yet, in times of mass extinction, terrestrial ecosystems also suffer species loss. Second, even within the marine realm itself, marine regression actually produces habitat expansion around oceanic islands, many of which harbor high species diversities today (Stanley 1987). Last, even given the highly provincial nature of marine biota today, Jablonski (1986a) argues that only 13% of modern families would become extinct if all the modern shelf biota were eliminated.

Further complications arise concerning the extent of continental inundation and regression. Given a sea level drop of equal magnitude, species–area effects could be expected to be more severe during intervals when the continents are covered with large epeiric seas than during intervals when absolute sea level stand is low and covers only narrow continental shelf zones. Thus the expected loss of species diversity due to the species–area effect is not a simple function of regression magnitude, but also depends on the preregression geographic condition of the marine biosphere.

A more promising causal link between marine regression and extinction may exist in climatic effects (Jablonski 1986a), though again the preregression configuration of land and sea is of major importance. Global climatic extremes may be buffered during intervals when the continents are covered with extensive shallow seas, which may simultaneously warm oceanic waters and dampen seasonal temperature fluctuations on the land. Marine regression would destroy this ameliorating effect, producing sharp climatic gradients between land and sea. Oceanic waters would become colder in the absence of warm-water influx from epicontinental seas, and seasonal temperature fluctuations on land would become extreme in the absence of the dampening effect of large shallow-water masses. An increase in exposed land area would heighten the albedo of the Earth and would reduce the carbon dioxide content of the atmosphere, because of increased weathering of exposed land mass. Both effects would produce overall global cooling, and cooling is consistent with the ecological signature of the Frasnian/Famennian mass extinction.

Interestingly, the Frasnian/Famennian crisis is one of the few known in the geologic record where a major biotic crisis occurs during an interval of global sea level *high stand* (figure 7.9). The maximum extent of continental inundation actually occurs in the late Frasnian. The overall rising sea level trend during the Frasnian is punctuated by several marine regressions, however, which are of shorter duration (figure 7.9). The sea level begins to fall in the Famennian, and the sea level drops seen in the latest Famennian are probably associated with the onset of continental glaciation in Gondwana, which is known to occur by the Middle *praesulcata* Zone (as discussed earlier).

Johnson (1974) suggests that a geologically rapid regressive–transgressive sea level pulse may occur during the latest Frasnian, which may have extinguished the "perched" faunas produced by the overall high stands of the Late Devonian oceans. In this scenario, Johnson (1974) stresses the preregression geographic configuration of Frasnian shallow seas, and does not simply stress the magnitude of sea level fall alone. A regression does appear to occur near the Frasnian/Famennian boundary (McGhee and Bayer 1985; Johnson et al. 1985; Sandberg et al. 1988), but many such sea level fluctuations occurred within the span of the Frasnian, none of which precipitated the massive disappearance of species seen during the latest phase of the Frasnian (McGhee 1982; Sutton and McGhee 1985; McGhee et al. 1991). More recent quantitative analyses of the fossil record have called into question the supposed correlation of sea level fall with any major biological event (McKinney and Oyen 1989; McGhee 1991, 1992; Valentine and Jablonski 1991).

Last, Johnson and colleagues (1985), using more refined correlations than were available to Johnson (1974), point out that the late Frasnian extinctions actually occur just before the end-Frasnian sea level fall. Thus neither regression-induced loss of habitat nor global cooling can explain the Frasnian/Famennian crisis, as the regression

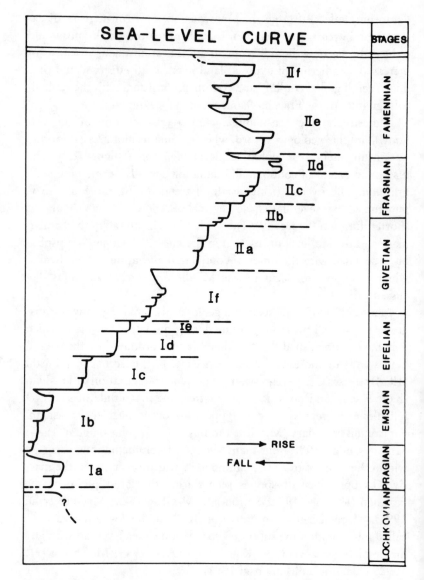

Figure 7.9. The Devonian sea level curve of Johnson et al. (1985). Note the maximum sea level highstand occurs in the Frasnian. (From Hallam 1992. Used with permission.)

postdates the extinction. Johnson and colleagues (1985) favor instead an extinction triggered by *marine transgression*. How? That scenario will be explored in the next section.

OCEANIC ANOXIA: POISON FROM THE DEEPS?

Periods of sea level high stand are generally viewed as producing equitable global climates favorable to both marine and terrestrial organisms. However, there are several proposals that the reverse might be true, at least for marine benthic ecosystems (Hallam 1989, 1992). Schlager (1981) argues that the extended Late Devonian transgression actually led to the "drowning" of major reef ecosystems, and House (1985) suggests that poisonous anoxic conditions are created with the spread of Frasnian shallow stagnant seas, conditions that would decimate bottom-dwelling marine organisms.

The Late Devonian oceans are very different from modern ones. Nutrient distribution is of great importance today, and whether eutrophic or oligotrophic, most water masses are fairly well oxygenated. Oceanic bottom waters throughout most of the Paleozoic are believed to be anoxic. Shallower waters on the continental shelfs and slopes are oxygen depleted from the Cambrian to the Carboniferous, resulting in widespread deposition of black shale facies during this interval of time (Berry and Wilde 1978). Anoxitropic biotopes, comprised of organisms (chiefly zooplankton) attracted to dysaerobic to anaerobic boundary zones, are widespread in Cambrian through Devonian seas (Berry et al. 1989). Only later in the Paleozoic do the oceans become more aerated, or oxygen "ventilated," and begin to resemble the water masses familiar to us today (Berry and Wilde 1978).

However, the upper 100 m or so of the oceans are generally well oxygenated (Berry et al. 1989). This is due to gas diffusion with the atmosphere, agitation of the seawater surface by wind action, and transport of oxygenated waters by shallow-water currents. It is in this region that the diverse benthic ecosystems flourish for most of the Devonian, prior to the Frasnian/Famennian mass extinction. Most of the organisms in the shallow benthos are aerobic, with only a few able to tolerate long periods of oxygen deprivation. Most of these same organisms are also sessile and unable to flee if covered with anoxic water masses. They simply smother.

In order to provide a potential killing mechanism the deeper anoxic waters must be brought out of the deeps and spread over previously oxygenated regions of the seafloor. One way to accomplish this is during marine transgression. First, with increasing water depths, a region of the seafloor that was previously above the oxygen minimum zone will now be covered by anoxic water. Second, marine transgression is often associated with warm water conditions, and warm water is less oxygen soluble, further raising the level of the oxygen minimum zone. Even in this scenario, however, organisms in less than 50 m of water depth may totally escape anoxia or may experience only temporary dysaerobic conditions, because of the aerating effects of wind and shallow currents (Wilde and Berry 1986).

It is clear that the Late Devonian is a time of widespread anoxic bottom waters, as the entire interval is characterized by black shale deposition in geographically extensive areas of the planet. The peculiar bituminous limestone and black shale units of the Kellwasser horizons can be found across Europe and northern Africa in the late Frasnian, demonstrating incursions of the oxygen minimum zone into very shallow water regions during this critical period of time (Buggisch 1972; Wendt 1988; Wendt and Belka 1991). In Europe, the spread of anoxic Kellwasser Limestone conditions clearly is associated with two transgressive events in the latest Frasnian (figure 7.10), and with a virtually continuous sea level high stand in Morocco well into the Famennian (Wendt and Belka 1991). At the same time, pelagic conodonts experience no restrictions in geographic distribution, and it appears they remain unaffected in the oxygenated upper-water column (Belka and Wendt 1992). Pelagic conodonts appear sensitive to surface-water temperature rather than to bottom-water anoxia (Geitgay 1985), and they are decimated in the late Frasnian.

Buggisch maintains the benthic anoxic effect is further augmented with high organic productivity in surface waters (see the section on climatic autocycles earlier), and notes that even where the black facies of the Kellwasser Limestones are not deposited, the characteristic positive excursion in $\delta^{13}C$ values still occurs (figure 7.10), indicating a shift in the total carbon reservoir of seawater (Joachimski and

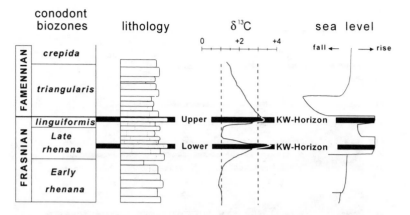

Figure 7.10. The twin black shale and bituminous limestone Kellwasser horizons ("KW"), which occur in the latest Frasnian in Europe and northern Africa. Both are associated with positive shifts in carbon isotope signatures and with sea level highstands. (From Joachimski and Buggisch 1993., Artwork courtesy of M. M. Joachimski.)

Buggisch 1993, 1994). Black shale facies around the world appear to be correlated with sea level rise in other parts of the Devonian as well (House 1985).

It is thus clear that anoxic conditions are associated with many broad shallow seas produced by transgressive phases in the Devonian. However, the entire Late Devonian is punctuated with transgressions and black shale tongues into shallow marine strata, and most of these events are not associated with major faunal perturbations (figure 7.11). Becker (1993), while advocating the eustatic rise/anoxic pulse model for the Late Devonian mass extinction, admits that such pulses occur throughout the Frasnian, and into the Famennian, without producing a biotic response similar to that seen at the Frasnian/Famennian boundary. He later advocates a "periodic 'greenhouse' climatic overheating episodes" model instead (Becker and House 1994; see "Thermal: Too Hot?" section preceding).

It is also not clear how the entire global marine ecosystem can become anoxic (Stanley 1987). As mentioned earlier, the upper 50 m of surface waters should always remain oxygenated, unless something drastic and unusual happens. Two such drastic scenarios have indeed been proposed: (1) oceanic overturn, and (2) asteroid impact. In the

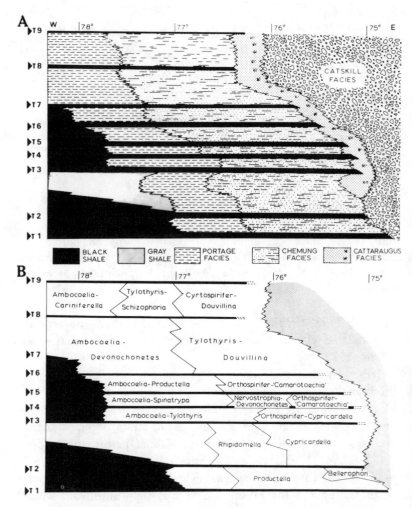

Figure 7.11. (A) The stratigraphic record of nine separate marine transgressions (arrowheads numbered T1 to T9), and associated black shale incursions into shallow water facies, during the Frasnian in New York State. (B) The distribution of benthic marine communities during this same interval of time, none of which experience significant biotic crises. (Modified from McGhee et al. 1991.)

oceanic overturn model, Wilde and Berry (1984, 1986) propose that the density structure of the ocean itself may become unstable when high-latitude waters approach 5°C during global cooling trends. The density stratification of the oceans breaks down (specifically because of the equalizing of the thermocline gradient), and bottom waters rise to the surface. Those bottom waters, in the Paleozoic, are anoxic and poisonous, containing sulfides, heavy metals, and other toxins. Even nutrients brought up by the overturn could ultimately be deadly, as they may trigger a "red tide" of overproductivity in the phytoplankton (Wilde and Berry 1984, 1986). Another way to mix oceanic waters thoroughly would be to impact the ocean with an asteroid (Geldsetzer et al. 1987). Both mechanisms have been proposed to explain the Frasnian/Famennian mass extinction.

Oceanic overturn would leave little physical evidence in the fossil record. The sharp positive excursions in $\delta^{13}C$ reported at some Frasnian/Famennian boundary sections around the world (figure 7.10, and further discussed in chapter 9) might be taken as evidence of the productivity bloom predicted by Wilde and Berry (1984, 1986) to accompany such an event. Anoxic poisoning is compatible with tropical ecosystem destruction, differential survival of deep-water marine species (presumably tolerant of anaerobic or at least dysaerobic environments), and differential survival of freshwater species (who never encounter the anoxic marine water). The phytoplankton and shallow benthos, above 50 m water depth, can be killed by anoxic poisoning only through an unusual event, such as the proposed mixing of marine waters all the way to the surface in an overturn event.

Thus many of the ecological signatures left by the Frasnian/Famennian mass extinction can be explained by catastrophic oceanic overturn; but this is not the case for two. True oceanic overturn should hit the high-latitude marine species as hard as the low, but this is not seen in the Frasnian/Famennian crisis (signal 2). And most important, there is the land. Terrestrial ecosystems are also affected by the Frasnian/Famennian crisis: the fish, the plants, perhaps even our amphibian ancestors (chapter 4). It is difficult to see how oceanic overturn, although perhaps producing a foul odor for a while, could have seriously affected terrestrial ecosystems.

ASTEROID IMPACT: DEATH FROM DEEP SPACE?

Perhaps the most interesting and controversial proposal made in paleobiology in the past decade is the proposal that the Cretaceous/ Tertiary mass extinction was triggered by the impact of a large asteroid with the Earth (Alvarez et al. 1980). There is an entire class of asteroid-sized bodies, the Apollo objects, that have orbits crossing that of the Earth and it is only a matter of time before a collision takes place. Thus it is entirely reasonable to propose that such an event has taken place in the past, and that the catastrophic results of such a collision could trigger global ecosystem collapse. In fact, as will be discussed in the next chapter, such a scenario was proposed specifically for the Frasnian/Famennian event by McLaren (1970) two decades ago, and though the suggestion was not taken seriously at the time, it was taken quite seriously a decade later (McGhee 1981a; McGhee et al. 1981; McLaren 1982, 1983).

A problem with the impact hypothesis is the geologically instantaneous, but short-term, effect of such a collision. The Frasnian/Famennian event was not instantaneous, in that multiple extinction events and diversity crises occurred in the latest Frasnian and earliest Famennian, spanning an interval of time up to 3.0 Ma in duration (as discussed in detail in chapter 3). This does not rule out asteroidal or cometary impact as the cause of some of the extinctions, but it does indicate that a single impact is not sufficient in and of itself to explain the multiple periods of elevated extinction rates. Alternatively, it could be proposed that the multiple periods of high extinction in the Frasnian were triggered by *multiple impacts* distributed over a geologically significant period of time, in a sort of "asteroidal shower." The multiple impacts hypothesis will be explored further in chapter 8.

It was the apparent nature of the biological evidence alone, however, that inspired McLaren's impact hypothesis. McLaren (1970) argued that the Frasnian/Famennian extinction was simply "too sudden to allow any but a catastrophic interpretation." If catastrophic, what triggered the catastrophe? Impressed by an earlier paper by Dietz (1961) on the possible consequences of a large meteorite impact, McLaren (1970) invoked an oceanic impact-generated tidal wave, perhaps 6 or 7 km high, as a kill mechanism. We now know, almost a quarter of a century later, that the global consequences of a large bolide impact are catastrophic indeed, with tsunamis playing only a small part in the overall environmental disaster. What would happen if the Earth were impacted by a large asteroid, one with a diameter of 10 km or so?

Direct Blast Effects?

Obviously, all life in the target area would be either vaporized at the impact site or killed by immediate blast effects in a lethal radius around the resultant crater. The largest human-made explosion was the 58 megaton bomb blast produced by the former Soviet Union in 1961 (Jones and Kodis 1982). A 1 megaton bomb explosion releases around 4.2×10^{22} ergs of energy; thus the Soviet fusion bomb released around 250×10^{22} ergs. In comparison, the impact of the Cretaceous/Tertiary asteroid with the Earth is estimated to have released over 10^{29} ergs, that is, $10,000,000 \times 10^{22}$ ergs, or 2.5 million megatons of energy (Jones and Kodis 1982). A large explosion.

Raup (1982) has demonstrated that mere megatonnage is not enough, however. The Earth's biosphere is sufficiently robust to withstand very large "lethal radii" (up to an entire hemisphere of the planet) without producing the diversity losses seen to occur in geologic time at the mass extinction horizons. Raup's research actually demonstrated the surprising robustness of the Earth's biosphere back in 1976, but he delayed publication of those results for over half a decade. The story behind that decision will be told in chapter 8.

Indirect Blast Effects?

Indirect blast effects include impact-generated earthquakes, and wildfires ignited by direct radiation from the blast fireball and ballistic projection of molten debris from the crater site. If ejected suborbitally, molten material will rain down at great distances from the target area. Indirect effects may be felt far outside the actual lethal radius of the crater impact site.

Shock waves propagating from the impact would trigger massive earthquakes on the land and tsunamis on the oceans. It is the latter event that most impressed McLaren back in 1969, because of its catastrophic effect on shallow marine shelf ecosystems. The shallow marine fauna comprise the bulk of the fossil record, because of their high preservation potential, and they are clearly decimated in mass extinction events. An impact-generated megatsunami could reach the opposite side of the Earth within 27 hours and still have a height of 150 m (O'Keefe and Ahrens 1982; Jansa 1983).

Wildfires are not to be ignored, for these are wildfires on a scale never known in human history. Wolbach and colleagues (1985, 1988) estimate that much of the land plant biomass of the continents was consumed by fire in the Cretaceous/Tertiary impact event.

Aftermath: Raining Acid and Poison?

Lewis and colleagues (1982) and Prinn and Fegley (1987) argue that a superheated plume, rising from the impact site, would heat the atmosphere to the point where nitrogen (the gas that makes up the major part of the Earth's atmosphere) would combine with oxygen to form nitric oxide (NO), which then would produce nitric acid (HNO_3). Acid rain, falling out of the sky in concentrations unknown in human history (a pH of 4 to 5 globally, and 0 to 1 near the impact site), could poison the upper surface waters of the oceanic photic zone, where the phytoplankton live. Even the calcareous shells of the zooplankton could be dissolved by the acid, which would have devastating effects on land plants as well.

This scenario, like all involved in large-body-impact studies, is highly model dependent. Zahnle (1990) suggests that a much-scaled-down version of the Prinn and Fegley (1987) predictions may be closer to reality.

In addition to nitric acid rain, however, we have the poisonous effects of the wildfires mentioned earlier (Wolbach et al. 1990; Gilmour et al. 1990). Huge fires would produce dioxins and polynuclear aromatic hydrocarbons (PAH), which are toxic. The megawildfires at the Cretaceous/Tertiary boundary may have produced as much as 7 $\times 10^{16}$ g of soot, which would block sunlight and lead to initial cooling of the planet (see later). The effect of pumping enormous amounts of CO_2 into the atmosphere could then produce a post-impact greenhouse-induced rise in global temperatures by as much as 5° to 10°C (Gilmour et al. 1990; Wolbach et al. 1990).

Aftermath: Freezing in the Dark?

The global effect of an asteroid or comet impact on the Earth's total climate is difficult to predict. Clearly, large amounts of debris from the vaporized bolide and impact site would be injected into the atmosphere. Such a global dust cloud could result in the total blockage of sunlight from the Earth's surface. The planet might look from space like Venus today—shrouded in pearly white and gray bands of cloud, as light from the sun is reflected back into the darkness of space. The absence of sunlight below would result in the cessation of photosynthesis on the planet's surface, with resultant death of both the phytoplankton of the oceans and plant life on the land. In the absence of radiant energy from the sun, surface temperatures of the planet would quickly drop below freezing and animal life would die as well.

Alvarez and colleagues (1980) initially thought such a dust cloud might stay aloft for several years, resulting in virtual total elimination of even hardy plant life, which, unlike the phytoplankton, could regenerate itself from spores and seeds once sunlight broke through the dust cloud gloom. Atmospheric modeling by Toon and colleagues (1982) and Pollack and colleagues (1983) diminished these early estimates considerably, however. These studies concluded the aerosols generated by the impact would settle out too quickly to suppress photosynthesis for several years. The period of total darkness might last up to 6 months. Sunlight too weak for photosynthesis might last up to about a year. The oceans, being thermally inert, might cool down only a few degrees Celsius. Temperature on the continental landmasses, however, would drop rapidly below 0°C and remain below freezing for up to 2 years (Toon et al. 1982). Models are just that, however, and if you change the model you obtain different results. More recent modeling by Covey and colleagues (1990) suggests that this estimate of continental freezing (like the estimate of amount of acid rain earlier) might be excessive, and they have proposed a more scaled-down version of geographically patchy freezing and global temperature decline. The effect of megawildfire soot production, however, might well bring that estimate back up again (see earlier).

Global temperature decline is consistent with the ecological signature of the Frasnian/Famennian event, however; thus the impact hypothesis provides a possible forcing mechanism for the biological crisis. The chief difficulty with the hypothesis is its geologically short-term climatic effect, which is not consistent with the 3.0 Ma period of extinction events seen in the latest Frasnian and earliest Famennian (table 3.2). However, it should be noted that the climatic predictions of the impact hypothesis are strongly model dependent. The climatic effects of an asteroid or comet impact could be more long term than is currently predicted by most atmospheric models and may produce global changes on a timescale of tens of thousands, or perhaps millions, of years. Such long-term climatic effects would be more in concert with the pattern of species extinctions, and lack of new species originations, seen in the fossil record of the Frasnian/Famennian event (chapter 3).

In a recent study, Schultz and Gault (1990) suggest that an oblique impact on the Earth (rather than the often modeled vertical, or "fall out of the sky," impact) might trigger long-term climatic consequences. Their models indicate that a 10 km object, impacting the Earth at an oblique angle of 10° and a velocity of 20 km/s, could ricochet a debris spray of 100–1,000-m-sized fragments away from the impact site. The ricochet debris would exit the target site at hypervelocities and could be inserted into orbit around the Earth. From space, the Earth would appear like Venus in the first impact scenario discussed earlier; add to this picture a few dark rings similar to the darker bands of Saturn.

A ring of debris encircling the Earth would considerably prolong the climatic effects of the initial impact. First, the shadow of the ring on the Earth would provoke strong thermal gradients on the planet between those areas still receiving solar radiation and those not receiving it. Shielding part of the Earth's surface from solar radiation would also cut down on total energy received from the sun, and hence bring on planetary cooling. Second, the impact debris in orbit would fall back to the Earth over time. The orbital decay of the debris could take as long as 10 Ma, during which time the Earth would suffer the periodic impact of debris fragments 100 m to 1 km in diameter (Schultz and Gault 1990). The result would be a geologically significant time interval of massive climatic disruption and environmental stress.

The only other alternative for the impact scenario is to invoke a series of impacts, distributed over a 3.0 Ma time interval, in the latest Frasnian and earliest Famennian. The possibility of multiple impacts during the Late Devonian was suggested a decade ago (McGhee 1982), and this scenario is currently gaining more support, as will be discussed in chapter 8.

SUMMARY: THE KILLING MECHANISM

The single killing mechanism that satisfies all the ecological signatures of the Frasnian/Famennian mass extinction, outlined in chapter 6, is lethal temperature decline on a global scale. The second best runner-up is catastrophic oceanic overturn, but it fails to explain the loss of diversity in terrestrial ecosystems (signal 1; see chapter 6), and the differential survival of high-latitude marine faunas (signal 2).

Traditional global cooling scenarios such as glaciation, as discussed earlier, do not work for the Frasnian/Famennian crisis. The period of lethal temperature decline has to be geologically sudden, operating over 3.0 Ma at the very most (table 3.2). It is probably not a continuous 3.0 Ma temperature decline, but more likely occurs in two to five separate events in the latest Frasnian (figures 3.4 and 3.5). The episodic (series of discrete events) nature of the crisis may rule against the abrupt, but single, climatic collapse model of Crowley and North (1988).

Impact-produced global cooling fits the observed pattern of extinction only if (1) there is more than one impact (McGhee 1982) or (2) a single impact somehow produces global cooling that continues for a geologically significant period of time (the oblique impact model of Schultz and Gault 1990). The possibility that either of these two mechanisms actually triggered the Frasnian/Famennian mass extinction will be taken up in the next chapter.

VIII

Catastrophic Impact of an Asteroid or Comet?

THE IMPACT HYPOTHESIS

A Single Impact at the End of the Frasnian?

Many people these days are familiar with the Martian moon Phobos, captured in photographs by the U.S. Viking Orbiters in the 1970s. Although its name means *fear,* the poor moon itself seems to have taken quite a battering in its time (figure 8.1). An eerie stark object floating in the blackness of space, so very far away, it orbits a planet that has also captivated the imagination of generations of humans on the planet next closest to the sun. The inhabitants of the Earth, during the final days of the Cretaceous, might have had a more horrific reaction to such an object, for an asteroid, probably very similar in appearance to Phobos, was soon to fall from the sky—with catastrophic consequences.

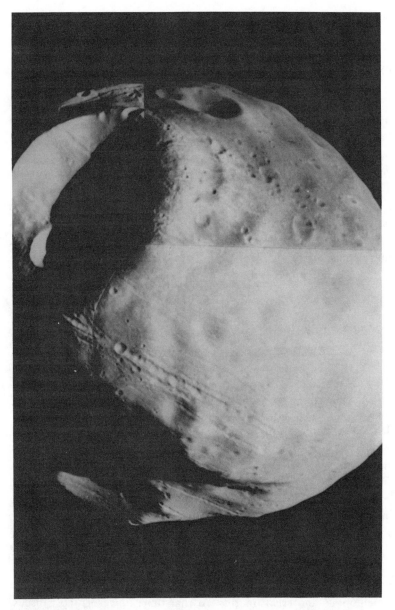

Figure 8.1. The Martian moon Phobos, photographed from 611 km by Viking Orbiter I. Phobos is most likely an asteroid, captured by Mars from the asteroid belt. The surface shown here (19 km by 22 km) is approximately twice the size of the asteroid initially proposed to have impacted the Earth at the end of the Cretaceous. (Photograph courtesy of NASA Team Project Leader Dr. Michael H. Carr and the National Space Science Data Center.)

Many people do not know, however, that the impact/extinction hypothesis was not originally proposed to explain the famous Cretaceous/Tertiary mass extinction. It was first proposed for the Late Devonian mass extinction. The hypothesis that the Late Devonian mass extinction was triggered by an asteroidal impact was put forth back in 1969 by the Canadian paleontologist Digby McLaren in his presidential address to the Paleontological Society and published one year later in the society's *Journal of Paleontology* (McLaren 1970). The idea was not taken seriously (as I have been told by older colleagues; I myself was still in high school at the time), but it was not totally ignored. Many thought McLaren was simply joking, as presidents are given free reign to discuss anything in their addresses, outrageous or not. Others were alarmed by McLaren's use of the phrase "catastrophic extinctions" and dismissed the hypothesis outright as a return to the then discredited nineteenth-century doctrine of catastrophism. In the middle 1970s the hypothesis was still being mentioned in the literature, though with a degree of bewilderment, as in this statement by a distinguished English paleontologist: "The general extinctions at the Frasnian/Famennian boundary, or close to it, have been discussed by McLaren (1970) and he, *doubtless with tongue in cheek,* suggested a cataclysmic explanation, invoking a meteorite setting up a giant and destructive tidal wave" (House 1975:474; emphasis mine).

During 1974 and 1975 I was engaged in Devonian paleontologic research as a master's degree student at the University of North Carolina at Chapel Hill (McGhee 1976). Although my interests were ecological, one of the spin-offs of that research was an attempt to locate the Frasnian/Famennian boundary in the stratigraphic sections I was studying (McGhee 1977; McGhee and Dennison 1980). The disappearance of the Frasnian marine fauna intrigued me as I hiked through the mountains of Virginia, West Virginia, and Maryland, trying to pinpoint the last horizon of some typical Frasnian species of brachiopods and cricoconarids. At that time, however, most discussions of the Frasnian/Famennian event centered around the Johnson (1974) "extinction of perched faunas" marine regression model (see chapter 7).

In 1975 I began doctoral studies at the University of Rochester, eventually studying theoretical morphology under the direction of

David M. Raup. (An aside: for a brief glimpse into graduate student life during those years at the University of Rochester, see "The Man Who Studied Horseshoe Crabs" in Ward 1992:139–141. During this same period at Rochester Jack Sepkoski was quietly working away on his now famous Compendium of fossil families; see Sepkoski 1994). In a classic case of chance and coincidence, Dave Raup was also interested in the Frasnian/Famennian extinction, and moreover took the McLaren impact hypothesis very seriously. Being of a probabilistic frame of mind, however, Raup was more impressed by a paper by Öpik (1951) concerning the expected frequency of collisions between planets and asteroidal or cometary bodies in the solar system. In general, the probability of small impacts on the Earth is quite high, but the expected frequency of collisions of very large objects with the Earth becomes vanishingly small, even given a Phanerozoic time span of several hundreds of millions of years. The question then becomes, "How large must a bolide impact be in order to trigger significant levels of extinction?"

To answer this question, Raup began a series of computer-simulated bombardments of the Earth. The bombs landed at random points on the Earth, and anything within a "lethal radius" of the bomb died. Obviously, one normally expects that the larger the "lethal radius," the more species will perish. This is not necessarily the case, however. Raup knew that a small direct hit on Australia or New Guinea would eliminate a large number of marsupial mammal species that were geographically confined, or endemic, to those areas and found nowhere else. Conversely, a large direct hit on an area inhabited by geographically widespread, or cosmopolitan, species might kill millions of individuals locally, but not the species themselves as they would have members that survived outside the "lethal radius." I spent the summer of 1976 as a research assistant to Raup, collecting modern biogeographic distributional data for his Earth bombardment project (see acknowledgments, Raup 1982). Most of his colleagues in the geological sciences department thought of the project as "one of Dave's crazier ideas," which would lead nowhere. Raup himself was aware of their assessment: "These catastrophic simulations were a source of much amusement for the graduate students and junior faculty at Rochester" (Raup 1986:44).

The results of the bombardment project were surprising. Raup discovered that establishing significantly large numbers of species extinctions could be done only by using very large "lethal radii" indeed—sometimes even as much as an entire hemisphere of the planet was required. The probability of bolide impacts of sufficient magnitude to produce such large "lethal radii" seemed vanishingly small and unlikely even on geologic time scales (at least for the past 600 Ma or so). I remember the seminar in which Raup presented these somewhat glum (for the impact hypothesis) conclusions. A biologist present at the talk was intrigued, however, and marveled at the apparent resiliency of the Earth's biosphere.

Raup thus concluded that a virtually global catastrophe was required to produced a mass extinction. Even given this apparently anti-"impact hypothesis" result, he hesitated to publish the results of the Earth bombardment project because he believed the research project itself "would be laughed at, or worse" (Raup 1986:45).

Raup overlooked one significant detail in the design of the bombardment project (and so did everyone else who discussed it with him at Rochester, for that matter): *the effect of fallout*. The climatic effects of fallout from the vaporized bolide and Earth at target site turned out to be the real killer, effects that could quickly become global, far from the lethal blast area. It was only after the publication of the now famous Alvarez and colleagues (1980) paper that Dave Raup finally published the results of his "crazy project" (Raup 1982).

It is ironic that the impact hypothesis was first taken seriously by the scientific community almost exactly one decade later than the McLaren (1970) address, but for a totally different mass extinction, the Cretaceous/Tertiary event, following the publication of the Alvarez and colleagues (1980) paper. Like McLaren (1970), the Alvarez and colleagues (1980) proposal was also immediately rejected by many geologists and paleontologists, without any serious examination or consideration, as a return to catastrophism. Unlike McLaren (1970), however, Alvarez and colleagues (1980) maintained they had *physical proof* of the impact: anomalous concentrations of the element iridium located exactly at the Cretaceous/Tertiary boundary in an Italian stratigraphic section. The significance of anomalous iridium concentrations, and of the additional types of physical evidence now being used as signatures of a bolide impact in the geologic record, will be discussed later. It is this key addition of the claim of physical proof of impact, and not just argumentation concerning the catastrophic nature of the biological evidence, that sparked serious examination of the impact hypothesis.

The search was on! The impact hypothesis was being hotly debated by scientists from all fields, but the hypothesis was now at least credible and tangible evidence could be sought: anomalous iridium concentrations near mass extinction horizons in the stratigraphic record. Of course, the huge majority of research effort was directed at the Cretaceous/Tertiary horizon, with its implications of a dinosaur killer. This was to be expected, as the mystery of the disappearance of the nonavian dinosaurs has puzzled paleontologists for over a century. Perhaps a catastrophic bolide impact would finally solve the persistent problem of the extinction of such a long-lived and successful group of animals.

I decided to take on the search for impact evidence at the Frasnian/Famennian horizon (McGhee 1981a; McGhee et al. 1981). After all, it was this event that was first proposed to have been triggered by an impact. An international group of over 110 scientists met in Snowbird, Utah, in late 1981 to discuss the impact hypothesis. It was an incredibly multidisciplinary event, with physicists, chemists, paleobiologists, ecologists, geologists, astronomers, climatologists, and more, all gathered together in what has now come to be known as "Snowbird I." I discussed my plans with Digby McLaren at that event and presented a short paper outlining the research to the conference. McLaren was very pleased that now, a little over a decade after he first proposed it, his idea was being taken seriously. He later wrote (McLaren 1982:482), in his paper for the conference proceedings concerning the Frasnian/Famennian crisis: "The size and the behaviour of this triggering mechanism is beyond my capacity to examine. I am happy to note, however, that an attempt is already being made to search for chemical evidence in a sedimentary sequence that embraces the horizon (McGhee 1981a,b)."

The search did not turn out to be a simple one, however. In fact, it unfolded in increasing complexity throughout the remainder of the 1980s and continues even now in the 1990s. The story of that search will be taken up in chapter 9.

Multiple Impacts During the Late Devonian?

Even at "Snowbird I" I began to have doubts that a single impact, however catastrophic, could explain the biological pattern of the Frasnian/Famennian mass extinction. For my paper in the conference proceedings, I outlined data indicating that the extinction was not a single instantaneous event (McGhee 1982). The ecological signature of the event pointed to global cooling as the proximal killing mechanism, but such cooling could not be due to a traditional glaciation scenario (or so I argued; see also chapter 7). If not glacial cooling, what? Climatic cooling, triggered by asteroidal impact, seemed to me a likely possibility (McGhee 1982).

How then to generate an extended period of global cooling by impact, rather than an instantaneous (but short duration, in geologic time) freezing following a single impact? To me, there seemed to be two possibilities:

1. A *single* (instantaneous) impact might trigger an *extended* (millions of years) climatic cooling phase on the Earth.
2. There could be *multiple impacts,* each triggering its own separate climatic cooling event, spread over a significant period of time.

Most climatologic models presented at "Snowbird I" called for short-term global refrigeration following an impact. More recent models, however, do suggest that a single impact might trigger global climatic effects spanning as much as 10 Ma (Schultz and Gault 1990).

At the time I concluded that a single impact would not produce the extended cooling event that seemed to be called for by the ecological data. I proposed, then, that the Earth could have been *"impacted by several large bodies during the Late Devonian"* rather than having been struck with a single large impact (McGhee 1982:498, see also report in Kerr 1993a:176).

Why not more than one impact? To some extent, this question occurred to me because of my experience with the Ries Crater in southern Germany. In 1977, as a doctoral student in theoretical morphology, I had traveled to the University of Tübingen to study with the school of constructional morphology at that university. During my stay of research, I traveled with colleagues to visit this 24 km crater at Nördlingen. It was indeed impressive, but there was more to the story. About 42 km to the southwest of the center of the Ries Crater is a much smaller structure (only 3.4 km in diameter), the Steinheim Crater. Both craters were the same age (14.8 Ma BP), and the falling objects clearly hit the Earth in a double punch.

If two impacts at once were possible, why not a shower of impacts scattered over time? In fact, even in 1981 cratering data existed, suggesting that the Late Devonian may have been an interval of time of increased impact frequencies (McGhee 1982:498).

In any event the time was not right for suggestions of multiple impacts, and the proposal was ignored or misunderstood. In a review of the "Snowbird I" conference proceedings, Van Valen (1984:130) writes: "[McGhee] reasonably relates the pattern to a cooling of the oceans but thinks the latter might be causally related to an impact. An impact-caused cooling should not last several million years, though." This, of course, was the problem that led me to suggest the possibility of multiple impacts in the first place.

Advancement of the Multiple Impacts Hypothesis

I originally suggested multiple impacts in the Late Devonian for two reasons: (1) to explain the presence of more than one single instantaneous extinction event during the time interval around the Frasnian/Famennian boundary and (2) to produce the extended global cooling that the ecological data seemed to demand (McGhee 1982; Kerr 1993a:176). Additional data supporting this hypothesis could be found in the terrestrial cratering record, even at the time of "Snowbird I" in 1981. Unfortunately, additional geochemical data were not to be forthcoming, as our initial search for hard physical evidence of impact turned up absolutely no iridium anomalies (McGhee et al. 1984, 1986a,b). The complex unfolding of the search for impact evidence at the Frasnian/Famennian boundary is taken up in chapter 9.

The concept of "stepwise mass extinction" did not enter into the debate back in 1981. Discussion at that time was between two antithetical extremes: "catastrophic mass extinction" and "graded mass extinction" (for a humorous review of the debate since "Snowbird I," see Flessa 1990). Supporters of the impact hypothesis maintained the view that all extinction occurred in a single instant, simultaneously planetwide. Biostratigraphers refer to this possibility as a "bedding plane event," a knife-edge mark that can be made in a stratigraphic section—except that here, in the case of a global mass extinction, that mark can be made in every stratigraphic section in the world, and moreover occurring precisely at the same horizon everywhere. Before the mark all species are present, then—wham!—above the mark all species are absent. Anti-impact supporters maintained that mass extinction, although it entailed a catastrophic ultimate loss of global species diversity, occurred over a measurable interval of geologic time. Over one to two million years, or perhaps even as much as ten million, the individual populations of species dwindled in numbers because of adverse environmental conditions. Sooner or later only one individual within a species was left, and when it died, that genetic lineage was gone forever. In the face of relentless environmental stress the species vanished, one by one.

By 1987 more and more physical evidence was accumulating to demonstrate that the Earth had indeed been impacted by a very large object at the biostratigraphic Cretaceous/Tertiary boundary. But the same biostratigraphic data simply did not match the expectations of the globally simultaneous bedding-plane event scenario. Some species seemed to vanish before the impact, whereas others appeared to survive the impact, only to succumb to extinction shortly thereafter. The pattern of species disappearances appeared to be "stepwise," with, however, the biggest "step" occurring at the impact horizon.

By 1984 other suggestions of the possibility of multiple impacts, scattered over a few millions of years but clustered around biological crisis horizons observed in the fossil record, began to appear in the literature (Clube and Napier 1984; Hut et al. 1985). Such an extended impact shower might be more consistent with the stepwise extinction pattern seen in the fossil record. Even here, however, the significance of clustered impacts *occurring over a significant period of geologic time* was overlooked, most people simply taking on the word *multiple* to mean that several impacts occurred all at once, and not just one big one. Thus when it was discovered that the Manson impact crater in Iowa was probably 9 Ma older than the Cretaceous/

Tertiary boundary and not the same age as the Chicxulub structure (Izett et al. 1993), the news was greeted with the report "New Crater Age Undercuts Killer Comets" stating that the fact that the two Late Cretaceous craters were of different ages "puts the swarm-of-comets idea back on the shelf" (see report of Kerr 1993b).

In 1988 a very specific proposal was made concerning Late Devonian multiple impacts: The Earth was impacted twice in the Frasnian, once during the Early *rhenana* Zone, and again during the *linguiformis* Zone (Sandberg et al. 1988). This proposal was based on the many years of laboratory and field experience that Charlie Sandberg and Willi Ziegler had amassed as conodont biostratigraphers. Using six primary Frasnian/Famennian control sections in Europe and North America, they and their colleagues argue that a stratigraphic sequence of twelve events can be found to occur everywhere on these two continents, and that they are probably global (table 8.1). These events occurred in the interval from the Early *rhenana* Zone to the Late *triangularis* Zone, spanning some 3.0 Ma of the latest Frasnian and earliest Famennian. The two impacts occur about 1.0 Ma apart, with the larger impact event positioned at the most severe pulse of the Frasnian/Famennian extinction, in the *linguiformis* Zone.

One of their control sections is the now famous Steinbruch Schmidt site in Germany, which we have met before in chapter 3 (figure 3.5) and will meet again in chapter 9. Their event-stratigraphic interpretation of the Upper Kellwasser horizon of this particular section is given in figure 8.2; compare it with the biological events outlined for this same section by Schindler (1990) in figure 3.5. Sandberg and colleagues (1988) believe the larger Frasnian impact occurred at the top of "Bed Number 16" in the Upper Kellwasser Limestone (figure 8.2). The top of this same bed is identified by Schindler (1990) as "Event C," where graphuroceratid ammonoids and entomozoan ostracodes disappear and a sharp reduction in homoctenid cricoconarids occurs (figure 3.5). Unfortunately, no iridium anomaly occurs in this bed, or elsewhere in the Steinbruch Schmidt section (McGhee et al. 1986a,b; Goodfellow et al. 1988), which could offer independent geochemical evidence of impact.

Europe and North America were united as a single continent in the Late Devonian (see chapter 5). The possibility exists that tectonic effects, unique to the Euramerican supercontinent, could underlie

Table 8.1. The Hypothetical Timing of Two Impacts and Other Significant Environmental Events in the Late Devonian.*

Time (years BP):	Sandberg et al. (1988) proposed events:
Early *crepida* Zone ca 365.5 Ma	
Late *triangularis* Zone ca 366.0 Ma	12. Onset of continental glaciation in Gondwana, sea level falls in Euramerica.
Middle *triangularis* Zone ca 366.5 Ma	11. Marine transgression and re-establishment of marine faunas.
Early *triangularis* Zone ca 367.0 Ma FAMENNIAN	10. Climax of sea level fall, wave scouring of carbonate platform margin. 9. Continuation of marine transgression, telescoping of conodont biofacies.
FRASNIAN *linguiformis* Zone ca 367.5 Ma	8. IMPACT OF LARGE BOLIDE?, shock-induced tsunamis and storms. 7. Increased sea level fall and marine shallowing. 6. Severe marine regression begins. 5. Major marine transgression, anoxic bottom waters spread across most marine basins.
Late *rhenana* Zone ca 368.0 Ma	4. Re-establishment of many marine faunas, but no reefs.
Early *rhenana* Zone ca 368.5 Ma	3. IMPACT OF SMALL BOLIDE? (or close pass by to Earth by large bolide?) 2. Marine regression. 1. Marine transgression and "drowning" of Frasnian reef framebuilding organisms.

* Based upon event stratigraphic analysis of numerous European and North American Frasnian/Famennian boundary sequences, proposed by Sandberg et al. (1988).

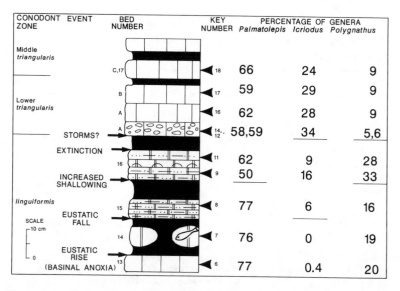

CONODONT ZONE	EVENT	BED NUMBER	KEY NUMBER	PERCENTAGE OF GENERA		
				Palmatolepis	*Icriodus*	*Polygnathus*
Middle *triangularis*						
		C,17	18	66	24	9
Lower *triangularis*		B	17	59	29	9
		A	16	62	28	9
	STORMS?	A	14, 12	58,59	34	5,6
	EXTINCTION	16	11	62	9	28
	INCREASED SHALLOWING		9	50	16	33
linguiformis		15	8	77	6	16
SCALE	EUSTATIC FALL					
10 cm		14	7	76	0	19
0	EUSTATIC RISE (BASINAL ANOXIA)	13	6	77	0.4	20

Figure 8.2. The event stratigraphic interpretation (left half of figure) of Sandberg et al. (1988) for the Upper Kellwasser horizon (*linguiformis* Zone) at Steinbruch Schmidt, Germany. Compare this section interpretation with the hypothesized sequence of events given in table 8.1, and the actual biological events given in figure 3.5. Right half of figure gives proportional abundance of three conodont genera in select beds ("key number"). Stratigraphic changes in water depth are reflected by percentage changes in *Palmatolepis-Polygnathus* (deep water) versus *Icriodus* (shallow water) in the conodont fauna. (Reprinted from Sandberg, Ziegler, et al. (1988), *Courier Forschungsinstitut Senckenberg* 102, figure 7. With permission of the authors.)

some of the events found by Sandberg and colleagues (1988); that is, they might not all be global. Hou and colleagues (1992) are conducting similar event-stratigraphic analyses of their sections in China, which is a separate continent and in a totally different part of the world than Euramerica in the Late Devonian. They also propose that two impacts occurred on the Earth in the late Frasnian— but they are not the same two proposed by Sandberg and colleagues (1988)!

Hou and colleagues (1992) back up their proposed two-impact events with additional geochemical evidence: iridium anomalies (table 8.2). However, *both* impacts are proposed to occur in the *linguiformis* Zone, and evidence of a yet older impact in the Early *rhenana* Zone

Table 8.2. The Hypothetical Timing of Two Impacts Within the *linguiformis* Zone, and Other Environmental Events Before and After These Impacts*

Time (years BP):	Hou et al. (1992) proposed events:
Early *crepida* Zone ca 365.5 Ma	
Late *triangularis* Zone ca 366.0 Ma	
Middle *triangularis* Zone ca 366.5 Ma	7. Marine transgressive event. Widespread nodular limestone unit with palmatolepid biofacies.
Early *triangularis* Zone ca 367.0 Ma FAMENNIAN	6. Storm event. Bioclastic grainstone with mixed palmatolepid and icriodid biofacies.
FRASNIAN	
	5. BOLIDE IMPACT? An iridium anomaly of 226 pg/g in a mudstone barren of fossils but with a sharp negative shift in $\delta^{13}C$, at the very top of the zone.
	4. Marine anoxic event. Black shale and nodular limestone unit rich in carbon and pyrite.
linguiformis Zone ca 367.5 Ma	3. BOLIDE IMPACT? An iridium anomaly of 316 pg/g in a bed of black shale.
Late *rhenana* Zone ca 368.0 Ma	2. Marine regressive event.
	1. Earthquake, collapse, or submarine landslide event. Deepwater limestone breccia unit extending 300 km along fault margin between deep basin and shallow platform.
Early *rhenana* Zone ca 368.5 Ma	

* Proposed by Hou and colleagues (1992).

is lacking. We will return to China and examine these sites in detail in chapter 9.

Could three impacts have occurred during the late Frasnian? Is there any hard physical evidence to support these various impact scenarios—undisputable evidence of impact, that everyone can agree on? The answer is yes, and that answer is *impact craters.*

SMOKING GUNS: LATE DEVONIAN IMPACT CRATERS

One obvious physical sign of impact is the existence of the crater produced by the collision of the bolide with the Earth. Interestingly, the Late Devonian does appear to have been a time when several large objects impacted the Earth; several probable craters exist that are generally dated as late Frasnian to early Famennian in age (McGhee 1982:498). One structure, the Siljan Ring in Sweden, is presently dated at almost exactly the Frasnian/Famennian boundary and is thus a likely candidate for a single-impact extinction scenario, but the existence of several Late Devonian craters might also be used to support the extended shower of impacts scenario outlined earlier. The choice between the one-impact or multiple-impact scenario hinges on two key factors: the nature of the biological data and the ages of the craters (or other physical evidence of impact, as will be discussed in chapter 9). The complex nature of the biological data has been discussed earlier in this chapter and in chapters 3 and 4. What do the crater ages now have to say?

Six impact craters are currently dated (Grieve and Robertson 1987) as possibly having been produced by impacts during the time range of the Late Devonian, from 377.4 to 362.5 Ma BP (Harland et al. 1989 timescale), and are listed in table 8.3. Interestingly, five of the six craters are also listed as potential Late Devonian impacts in the earlier study of Grieve and Robertson (1979), but their ages are not all the same. That is not surprising, as new data become available with time, old errors are corrected, and science progresses—sometimes painfully slowly, particularly when an immediate answer is desired.

Contrasting the A list with the B list in table 8.3 is instructive. We can see where progress has been made, and where not (meaning work

Table 8.3. Known Impact Craters with Age Ranges that Span the Frasnian/
Famennian Boundary (367.0 Ma BP)*

Crater	Diameter (km)	Age (years BP)
A. *After Grieve and Robertson (1987):*		
Siljan, Sweden	52	368 ± 1
Charlevoix, Quebec, Canada	46	360 ± 25
Kaluga, USSR	15	380 ± 10
Lac La Moinerie, Quebec, Canada	8	400 ± 50
Crooked Creek, Missouri, USA	5.6	320 ± 80
Flynn Creek, Tennessee, USA	3.8	360 ± 20
B. *After Grieve and Robertson (1979):*		
Siljan, Sweden	52	365 ± 7
Charlevoix, Quebec, Canada	46	360 ± 25
Kaluga, USSR	15	360 ± 10
Crooked Creek, Missouri, USA	5.6	320 ± 80
Flynn Creek, Tennessee USA	3.8	360 ± 20

* Craters ranked in order of diminishing size. From Grieve and Robertson (1979, 1987).

is needed). Clearly, no progress has been made concerning the ages of the Charlevoix, Flynn Creek, and Crooked Creek craters, as their age determinations have not changed and they still have substantial error ranges. The Kaluga Crater is reported to be older in the A list relative to the B, but still with an upper range that falls within the Frasnian. A new, but poorly dated, crater has been added since 1979 to the A list: the Lac La Moinerie.

How big must the impacting body be in order to trigger a global ecosystem collapse? There is no definitive answer to this question at present, but models do exist and they are constantly being refined (Raup 1992a,b). Raup (1992a) has attempted to construct a mathematical function for the relationship between crater diameter size and expected species kill (figure 8.3). If we use the empirical 70% species kill figure (see table 2.1) for the Frasnian/Famennian extinction as seen in the eastern North America data (McGhee 1982), a crater diameter of around 150 km is predicted (figure 8.3). If we use the higher estimate of 82% species kill (table 2.1) from the Jablonski (1991) model, a crater diameter of 250 km is required.

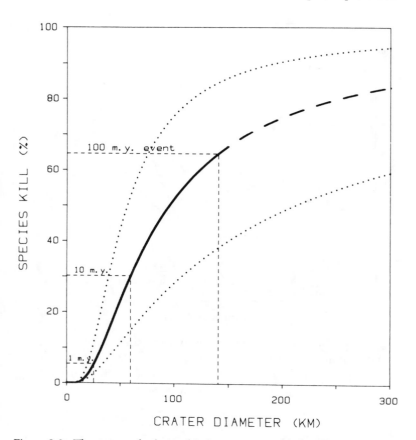

Figure 8.3. The proposed relationship between species kill and impact crater diameter, from the model of Raup (1992a). (Used with permission of the author.)

Jansa and colleagues (1990) and Jansa (1993) have also considered the question of threshold constraints for impact-induced mass extinction and have derived a somewhat different kill curve (figure 8.4). Jansa prefers a hyperbolic curve, rather than the sigmoidal curve of Raup, because of Jansa's belief that "additional impacts contributed to environmental stresses at the K/T boundary" (Jansa 1993:280). If due to a single impact, the 50% to 60% generic kill of the Frasnian/Famennian mass extinction (table 2.1) would require a crater diameter in excess of 300 km, a crater larger than the Cretaceous/Tertiary Chicxulub structure, in the estimation of Jansa (1993).

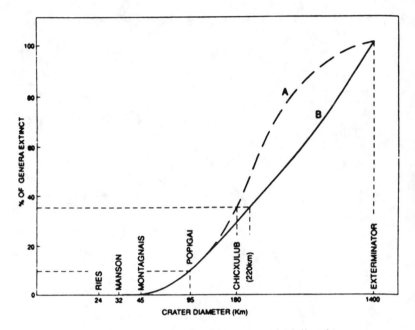

Figure 8.4. The proposed relationship between species kill and impact crater diameter, from the model of Jansa (1993). Dashed curve A is the sigmoid function of Raup (1992a), solid curve B is hyperbolic function preferred by Jansa (1993). (From Jansa 1993. Used with permission of the author.)

The largest known crater from the Late Devonian is the Siljan, at 52 km (table 8.3). Thus even this large crater is only one-third the size predicted from the Raup (1992a) model to be necessary to produce a 70% species kill. It is only one-fifth the size predicted to be necessary to produce an 82% species kill.

What would be the result, however, of the impact of three bolides of sufficient size to produce a 50 km crater each, rather than one bolide big enough to produce a 150 km crater? Also, what would be the result if those three impacts occurred spaced over one to two million years, and not all at once? The biological data cannot be explained by a single impact, or at least not all the data; thus seeking a single large crater may not even be required or expected.

There may well have been an impact in the Late Devonian large enough to produce a 150 km crater, say a crater in the ocean floor that was long ago destroyed by the tectonics of the constantly shifting plates of the Earth's outer surface. The crater age data, however, do show us the existence of several smaller craters of possible Late Devonian impacts. And they are not hypothetical; they are real—but are they Late Devonian?

The Siljan Crater

The Siljan crater has seen a marked improvement in $^{40}Ar/^{39}Ar$ age determination (table 8.3). The impact producing the Siljan structure is now considered to have occurred between 369 and 367 Ma BP, or somewhere from the base of the *jamieae* Zone to the base of the Early *triangularis* Zone in the Frasnian (Ziegler and Sandberg 1990; see also table 1.6). The latter horizon is the formal Frasnian/Famennian boundary (Cowie et al. 1989), but the age of the base of the Early *triangularis* Zone is 367.0 Ma BP only if the Harland and colleagues (1989) timescale is correct (see chapter 1).

A great deal of the new information concerning the Siljan crater is due to the choice of this structure by the Swedish Power Board as a deep-drilling site. Deep drilling was undertaken in the crater to test the "deep-earth methane hypothesis" (Gold and Soter 1980), namely, that there might exist economically significant reservoirs of methane within the Earth that are mantle derived. Hypothetically, fracture of the Earth's crust by a large impact would produce deep fissures through which this theorized methane could escape to the surface. The Swedish Power Board did not find any methane. Scientific study of the crater itself benefited greatly, however (Komor et al. 1988).

The Siljan impact event, needless to say, is the number one candidate for impact-generated mass extinction at the Frasnian/Famennian boundary.

The Flynn Creek Crater

The Flynn Creek Crater is particularly illustrative of the urgent need for multidisciplinary effort in impact crater dating. The structure is listed with a potential age spread from 380 to 340 Ma BP in table 8.3, thus with an impact supposedly occurring between the early

Givetian Stage of the Middle Devonian (380.8–377.4 Ma BP) to the middle Holkerian Stage of the Early Carboniferous (342.8–339.4 Ma BP, Harland et al. 1989 timescale). However, the age of this crater can be further constrained using biostratigraphic data. Within the crater is a bedded breccia deposit, which is the oldest unit to have infilled the structure after the initial impact (Huddle 1963). The original impact occurred in the Appalachian Sea, and the bedded breccia unit was deposited as a submarine stratum within the void left in the epicontinental seafloor by the blast. Fortunately, fossils of the conodont species alive at the time are also found in the breccia unit. They are dated to the Lower *Polygnathus asymmetricus* Zone, and the breccia is considered of equivalent age to the Genundewa Limestone Member of the Genesee Formation in New York State (Huddle 1963). The Lower *asymmetricus* Zone has been chosen to define the base of the Frasnian (Ziegler and Klapper 1985), and the Genundewa Limestone in New York State has likewise been shown to be of earliest Frasnian age in subsequent study (Kirchgasser et al. 1988). In the Late Devonian Standard Conodont Zonation (Ziegler and Sandberg 1990), the former Lower *asymmetricus* Zone is now split into the Late *falsiovalis* Zone and the *transitans* Zone (table 1.6). The base of the ammonoid *Manticoceras* Zone falls within the conodont *transitans* Zone (Ziegler and Sandberg 1990), and Kirchgasser and colleagues (1988) show the Genundewa Limestone as occurring at the level of the base of the *Manticoceras* Zone; hence the age of the breccia unit within the Flynn Creek Crater is of the *transitans* Zone.

What does this tell us? First, it rules out most of the range in age for the crater obtained by radiometric dating, namely, from early Givetian to Early Carboniferous (380–340 Ma BP, table 8.3). With conodonts from the early Frasnian *transitans* Zone present in a breccia unit deposited after impact, the crater cannot be any younger than early Frasnian. Second, depending upon how quickly the breccia unit was deposited in the crater following the impact blast, the crater is probably no older than late Givetian to early Frasnian. [Grieve and Robertson (1987) would agree that it is no older than early Givetian.] In conclusion, the Flynn Creek impact occurred very near the Givetian/Frasnian boundary, or around 377.4 Ma BP (Harland et al. 1989 timescale).

The Charlevoix Crater

The Charlevoix Crater is listed with a potential age spread from 385 to 335 Ma BP. Thus this impact may have occurred sometime between the middle of the Eifelian Stage of the Middle Devonian (386.0–380.8 Ma BP) to the early Brigantian Stage of the Early Carboniferous (336.0–332.9 Ma BP; Harland et al. 1989 timescale), to the assessment of Grieve and Robertson (1987). That assessment may soon change, as we shall discuss.

The Charlevoix Crater, originally called the La Malbaie structure, was discovered by Jehan Rondot during his geologic mapping for the Quebec Department of Natural Resources. His work drew the attention of Meteoritical Society researchers at the Dominion Observatory of Ottawa, Canada, who confirmed the impact origin of the structure (Robertson 1968). Since its discovery, the crater has been extensively studied by Rondot (1970, 1971, 1975, 1979). He estimates the size of the impacting body to have been roughly 2 km in diameter, impacting with a velocity of 20 km/s and an energy release of 2.5×10^{28} ergs. The result is a 46 km diameter crater in Quebec, half of which is now under the St. Lawrence River.

Most important, Rondot (1971) has found impactite preserved in a few small sites within the crater. Impactites are rocks that melted during the impact, and they are composed of mixed debris and melt glass. In so ancient a structure, most of such material has long been eroded away and destroyed. The Charlevoix impactites are found only in a region that did not undergo a major vertical displacement following the impact, approximately equidistant from the center of the crater and the inner ring, and that are preserved in a small depression that escaped erosion.

I was contacted by Jehan Rondot following "Snowbird I," as he had learned that Ed Olsen (Field Museum of Natural History), Carl Orth (Los Alamos National Laboratory), and I (at Rutgers) were conducting searches for iridium anomalies in New York State and Europe (see chapter 9). He kindly sent us a sample of the impact for iridium analysis. We found nothing. Iridium was below the detection limit of INAA (instrumental neutron activation analysis), that is, in concentrations of less than 1 ng/g. Chromium concentrations were at

38 ng/g, and cobalt at 21 ng/g, in the sample we ran; that is, nothing unusual was found. Iridium may have been present in the impacting bolide, but deposited as fallout well away from the central crater. More likely, the impacting body was not rich in iridium to begin with. Two samples of the impactite have been K/Ar dated at 372 Ma BP and 342 Ma BP (Rondot, 1971). Those dates would suggest an impact in either the middle Frasnian or the Early Carboniferous. The average of those two data is 357 Ma BP, or the earliest Carboniferous, very near the Devonian/Carboniferous boundary, according to the time scale of Harland and colleagues (1989).

It may be possible to set limits on the age of the Charlevoix impact still further by using biostratigraphic data, as demonstrated for the Flynn Creek Crater in the preceding discussion. A very peculiar sedimentary unit—the Spechty Kopf Formation—exists in the Appalachian Mountains, which can be traced for over 400 km and which has puzzled sedimentologists for years. It is called a "polymictic diamictite," which is sedimentologic jargon for a rock that looks very much like the poorly sorted mixture of everything from boulders to silts found in a glacial till, but it is probably not a glacial till. The central Appalachians, where this peculiar stratum is found, was located in the tropics during the Late Devonian (McGhee and Sutton 1981). Temperatures were too high for extensive glaciation, even in the highest mountains. Yet the "polymictic diamictite" is composed of boulders as big as 0.5 m, a chaotic mixture of gravels and exotic cobbles from variable terranes, grading slowly upward to finer pebbles, sands, and silts at the top of the unit.

It has long been thought that this peculiar unit is of latest Devonian age, very close to the Devonian/Carboniferous boundary. The coincidence of this peculiar sedimentary stratum and the Charlevoix impact crater has not gone unrecognized. Woodrow and colleagues (1990) have proposed that the unit was the result of an impact explosion, being produced in the central Appalachians by the earthquake-style initial blast shock, and subsequent aftershocks, following an impact. In addition to the peculiar sedimentology of the unit itself, preliminary evidence indicates that shocked minerals may be present (possible independent evidence of impact, as discussed later). Woodrow and colleagues (1990) consider the impact to have occurred either in the Appalachian Sea to the west (i.e., Flynn Creek Crater) or the Acadian Mountains to the north (i.e., Charlevoix Crater). The Flynn

Creek Crater can be ruled out, as it is too old, as discussed earlier. The Charlevoix Crater is becoming a more and more likely candidate, however.

John Richardson (of the British Museum of Natural History) has found palynomorphs within the Spechty Kopf "polymictic diamictite" that indicate that deposition occurred rapidly and all at once in a unit over 400 km (Woodrow et al. 1990). The unit falls in the paleobotanical *Lepidophyta* Zone, indicating a time position in the latest Famennian, very close to the Devonian/Carboniferous boundary. Correlation between the paleobotanical zonation of terrestrial areas and the paleozoological zonation of marine areas is still imprecise, or not as precise as many would desire. At present, the *Lepidophyta* Zone correlates with the *praesulcata* zones, but precisely which (Early, Middle, Late, or a combination of the three) is not known at present (Richardson and McGregor 1986; Richardson and Ahmed 1988).

Thus, if (and there are a lot of "ifs" in the reasoning that follows) the "polymictic diamictite" in the central Appalachians was produced by the shock and aftershocks of the impact event that produced the Charlevoix Crater, located just to the northwest in Quebec, and if current correlations between land-based and sea-based biostratigraphic zonations are correct, then the Charlevoix impact occurred in either the Early, Middle, or Late *praesulcata* Zone—that is, in the very latest Famennian.

The Kaluga Crater

The Kaluga crater now has an age range from 390 to 370 Ma BP, or from the early Emsian (390.4–386.0 Ma BP) to the middle Frasnian (377.4–367.0 Ma BP, Harland et al. 1989 timescale) within the Devonian. Thus an impact occurred within the Devonian, but we cannot say exactly when. It may have been in the Frasnian, but it may not. And that, unfortunately, is about all we can say at present.

Crooked Creek, Lac La Moinerie, and Other Craters

The Crooked Creek Crater has an age range from 400 to 240 Ma BP, and the Lac La Moinerie an age range from 450 to 350 Ma BP (table

8.3). The former impact may have occurred anywhere from the Early Devonian to the Middle Triassic, and the latter anywhere from the Late Ordovician to the Early Carboniferous. These craters are so poorly constrained in time that it is not worth discussing them. That is to say, any discussion of them with regard to the Frasnian/Famennian extinction (or any other extinction within the time span, for that matter) is nothing more than speculation.

Other potential Late Devonian craters exist for which we currently have even less information. Lake Taihu in China is one we will encounter in chapter 9. The Kentland Dome in Indiana is another (Gutschick 1987). The peculiar Alamo Beccia, covering more than 7,500 km² in southeastern Nevada, has also been proposed to have been produced by a catastrophic impact (Warme et al. 1993; Warme 1994). As the Alamo Breccia occurs within the *punctata* Zone of the early Frasnian (Sandberg and Warme 1993), it may record the evidence of an impact that occurred only 0.5–1.0 Ma after the Flynn Creek impact in the very earliest Frasnian (see table 8.4). No doubt

Table 8.4. Range of Possible Zonal Positions for the Three Most Probable Late Devonian Impact Events*

Stage	Zone	Time Horizon of Impact Marked "*?"
TOURNAISIAN (Carboniferous)		*?
FAMENNIAN	Late *praesulcata*	*? Charlevoix Crater
	Middle *praesulcata*	*? 46 km diameter
	Early *praesulcata*	*?
	Late *expansa*	
	Middle *expansa*	
	Early *expansa*	
	Late *postera*	
	Early *postera*	
	Late *trachytera*	
	Early *trachytera*	
	Latest *marginifera*	

	Late *marginifera*	
	Early *marginifera*	
	Late *rhomboidea*	
	Early *rhomboidea*	
	Latest *crepida*	
	Late *crepida*	
	Middle *crepida*	
	Early *crepida*	
	Late *triangularis*	
	Middle *triangularis*	
FAMENNIAN	Early *triangularis*	* ?
FRASNIAN	*linguiformis*	* ? Siljan Crater
	Late *rhenana*	* ? 52 km diameter
	Early *rhenana*	* ?
	jamieae	* ?
	Late *hassi*	
	Early *hassi*	
	punctata	
	transitans	* ?
	Late *falsiovalis*	* ? Flynn Creek Crater
FRASNIAN	Early *falsiovalis* (pars)	* ? 3.8 km diameter
GIVETIAN	Early *falsiovalis* (pars)	* ?

* See text for data and argumentation.

even more evidence of Late Devonian impacts will be discovered in the future, now that the search is on.

Summary of the Smoking-Gun Evidence

If the data and argumentation developed earlier hold, we can now say the Late Devonian world was impacted by an extraterrestrial body at least three times: near the Givetian/Frasnian boundary, the Frasnian/Famennian boundary, and the Famennian/Tournaisian boundary (table 8.4). If so (another "if," and a big one), the coincidence of biostratigraphic "boundary" and impact "crater" is rather interesting (McGhee 1994). Is it a coincidence? Could the biostratigraphic stage boundaries of the Late Devonian be causally related to impact events?

The geographic distribution of the Late Devonian impacts discussed earlier is given in figure 8.5, shown on the paleogeographic reconstruction of Dalziel and colleagues (1994, see figure 5.12). Notice that they appear clustered; they all fall approximately on a line that could be drawn from the eastern United States up through eastern Canada and across to and through northern Europe—that is, a line running from southwest to northeast, from 30°S to 30°N, along the eastern margin of Euramerica.

Can anything be made of this apparent line of impacts? Probably

Figure 8.5. Geographic distribution of the six known or potential Late Devonian impact sites. Note that the impact sites fall along a line trending southwest to northeast along the eastern margin of Euramerica. Solid circles position the three known Late Devonian impact sites, open circles with central points show the three additional potential impact sites in the Late Devonian, see text for discussion. Paleogeographic reconstruction after Dalziel et al. (1994), crater positions from Grieve and Robertson (1987).

not. It is possible that this line represents a great circle intersect along which the impacting objects streaked in from outer space during the Late Devonian, all coming in—one after another—in a plane obliquely oriented to the plane of the ecliptic of the solar system. It is just as possible that the line simply is an artifact of sampling. There are more geologists, and research universities, in North America and Europe than in many other areas of the world. If one takes the Grieve and Robertson (1987) map of known impact structures, tacks it to a wall, steps back a few paces, and simply concentrates on the dots representing craters, a pattern emerges. There are lots of craters in North America, western Europe, and (curiously) Australia. There are very few craters in South America, Africa, and the large expanse of Asia and Siberia. Does this mean that North America and Europe are preferentially hit by objects from outer space, for some reason? No, it probably simply reflects the fact that more scientists living in these areas of the world are interested in hunting for impact structures. Vegetation is probably also a big factor. The Earth is covered in dense vegetation in (what remains of) the tropical rainforests of South America, Africa, and Asia. Craters probably exist there, but will not be found until humans have succeeded in cutting down and destroying all the trees. The factor of visibility probably also explains the cluster of craters seen in western Australia; this region is desert, with little vegetative cover, and many structures are even visible from the air.

No one knows how many other Late Devonian impacts occurred of which no evidence remains. The Devonian world, like the modern, was mostly covered by oceanic waters. (The name *Earth* is a real misnomer given to the planet, of course, by land-based animals.) However, those Devonian ocean basins are long gone, recycled into the Earth's interior along plate subduction zones. We can still find Jurassic or younger craters produced by oceanic impact in the existent seafloor. We cannot do so for the Devonian, and the only potential record we have left to determine the frequency of impact during the Late Devonian is the small sample of impacts that struck the continents, and the even smaller sample of impact craters that have survived erosion and have been discovered today.

With such a meager sample, it even seems certain the Earth was impacted at least three times within the Late Devonian (table 8.4).

Those are the *preserved* hits, and two are quite large (craters around 50 km in diameter, see table 8.3). The *actual number of hits* may have been much larger.

I began this section with a question: What do the crater ages have to say? I would argue, and have done so before, that they say *multiple impacts*.

OTHER EVIDENCES OF IMPACT

Before 1980 about the only widely accepted evidence for large-body impact on the Earth was the physical presence of a crater. This situation has changed dramatically in the past fourteen years, following the publication in 1980 of the Alvarez team's hypothesis that the Cretaceous/Tertiary mass extinction was triggered by the impact of an asteroid with the Earth. In the ensuing fourteen years, many new types of data have been amassed and argued to be additional evidence of impact (chiefly with respect to the Cretaceous/Tertiary mass extinction). These new classes of data are: (1) iridium anomalies, (2) shock-metamorphosed minerals, (3) isotopic ratio comparisons, (4) anomalous carbon (soot) layers, (5) stable isotope anomalies, and (6) impact melt glass (tektites). Some of these new types of data have been argued to be direct evidence of impact, and some indirect. All have been disputed. I examine each proposed new line of evidence of impact here in turn, as they will be important in the next chapter, where the worldwide search for evidence of impact during the Frasnian/Famennian crisis is detailed.

Iridium Anomalies: Catastrophe from Below or Above?

The concept of an "iridium anomaly" was ushered into science by the now famous Cretaceous/Tertiary study of Alvarez and colleagues (1980). They found concentrations of iridium of around 7 ng/g, almost an order of magnitude above background concentrations, in stratigraphic sections in Italy directly at the Cretaceous/Tertiary boundary. Because iridium is depleted in the Earth's crust, relative to its known abundance in the solar system, they proposed an extraterrestrial source—a 10-km-diameter one. The implication that the dinosaurs may have been wiped out by an asteroid falling from the sky was crystal clear, and an uproar of debate ensued.

The debate concerning the significance of anomalous concentrations of iridium in sediments has raged for more than a decade now, and any attempt to summarize all the data, interpretations, and arguments presented in these last fourteen years could easily fill a book in itself. Therefore I shall not attempt an extensive overview; instead I shall endeavor to be succinct. There appear to be two possible sources for the anomalous iridium: from extraterrestrial bodies outside of the Earth and from deep within the Earth itself. Certain classes of extraterrestrial bodies (such as chondritic meteorites) are known to be rich in iridium. Thus the vaporization of an iridium-rich body on impact with the Earth is clearly one source. Such an impact would produce an atmospheric dust cloud that would envelope the entire Earth, with iridium fallout occurring globally. The climatic and biological effects of the generation of such a global dust cloud would be catastrophic (as discussed in chapter 7).

The other potential source is deep in the mantle of the Earth. Deep plumes could tap this source of iridium and spew it forth at the Earth's surface in periods of volcanism whose magnitude and intensity are unknown in human history, but do occur in geologic time. Volcanic dust, poisonous gases, and iridium would envelope the planet, and fallout would occur globally.

Is not the latter scenario equally as catastrophic as the impact scenario? The principal difference between the two lies in timing: an instant in geologic time for catastrophic impact, an interval of geologic time for catastrophic volcanism. The problem of time resolution lies at the heart of the debate between the two.

It was long thought that if an anomalous concentration of iridium is found in a sedimentary sample, then it is due to an anomalous concentration of iridium being present in the environment during the time of deposition of the sediment, and we are back to the question of source discussed earlier. It is now known that this is not the case and that iridium can be artificially concentrated by a variety of mechanisms. This simply means that iridium anomalies now have to be analyzed a little more carefully before an interpretation is offered, for it now must be proved that the iridium anomaly is primary and not due to secondary enrichment (see Kyte 1988; Sawlowicz 1993).

Interestingly, it was the search for impact evidence at the Frasnian/ Famennian boundary that first demonstrated the fact that iridium could be biologically concentrated. That discovery was made back in 1984, and it will be discussed in chapter 9.

Shock-Metamorphosed Minerals: Mega-Explosions?

Quartz grains with characteristic shock lamellae were discovered at the Cretaceous/Tertiary boundary by Bohor and colleagues (1984), and both quartz and other minerals bearing shock lamellae have since been discovered at Cretaceous/Tertiary sites around the world. Minerals showing such features, where the lattice of the crystals bear planar deformations produced by shock, are known from nuclear bomb test sites and ancient impact crater deposits. Some shock metamorphism of minerals has been reported to have been produced by explosive volcanism, but the complex shock lamellae in Cretaceous/ Tertiary boundary samples appear to be characteristic of and unique to high-energy blasts. Care must be taken to rule out metamorphism induced by volcanism, but the presence of shocked minerals at a stratigraphic horizon is considered to be one of the best independent lines of evidence for bolide impact (Moses 1989).

Isotopic Ratios: Signatures of Extraterrestrial Visitors?

Since 1980 a series of studies have examined the ratio proportions of platinum group elements, or isotopes of platinum group elements, to see whether those ratios reflect cosmic elemental and isotopic proportions rather than terrestrial ones. The use of ratios, rather than absolute magnitudes, is tricky, as ratios are mathematically unstable. Small errors in absolute measurement can yield large errors in ratio proportions. Thus in such studies it is imperative to ensure that the signal measured is primary, and not that of secondary diagenetic alterations.

In an early study Luck and Turekian (1983) argued, from geochemical analyses for differential ratios of osmium isotopes, that the $^{186}Os/^{187}Os$ signature of the sediments at the Cretaceous/Tertiary boundary was not that of the normal crust and could be extraterrestrial in origin. They could not rule out a deep-mantle source, however; thus the use of osmium isotopic ratios encounters the same source problem as that for iridium anomalies (Moses 1989).

Ratios of ruthenium/iridium (Ru/Ir) and osmium/iridium (Os/Ir) are also being examined, as these elemental ratios have chondritic values that are quite different from terrestrial ones (Sawlowicz 1993).

Global Soot Layers: A Planet on Fire?

While conducting geochemical analyses of the Cretaceous/Tertiary boundary layer, Wolbach and colleagues (1985, 1988) noticed peculiar particles of carbon in the samples. The graphitic particles have the characteristic morphology of fluffy carbon aggregations produced by burning, or, to put it plainly, the morphology of soot. Finding the same peculiar carbon particles in samples from as far away as Denmark, Spain, and New Zealand, they proposed that the particles were indeed the soot deposits produced by burning, but burning on a global scale. In addition to soot, complex carbon fullerenes (C_{60}) have also been discovered at the Cretaceous/Tertiary boundary and are likewise argued to have been produced by wildfire (Heymann et al. 1994). Arguing that the Cretaceous/Tertiary boundary layer represents less than one year, the amount of carbon present in the layer would represent a global wildfire of enormous proportion, consuming a large part of the land plant biomass.

A global soot layer in and of itself is not direct evidence of a bolide impact. It does indicate a catastrophic event, and one consistent with wildfire ignition via radiation from the impact fireball and ballistic projection of molten ejecta far from the impact site.

Stable Isotope Anomalies and "Strangelove Oceans"

Stable isotope anomalies, in and of themselves, do not provide direct evidence for bolide impact. Their presence has been used instead as evidence of rapid and massive alterations in the physical environment of the Earth's atmosphere and hydrosphere, the catastrophic nature of which is consistent with the effects produced by the impact of a large body with the Earth (Bhandari 1991). The stable isotopes of three elements are of primary interest, and they will be discussed in this section. Those elements are (1) carbon, (2) oxygen, and (3) sulfur.

The stable isotopes of carbon are ^{12}C and the heavier ^{13}C, which has an additional neutron. (Carbon has a still heavier isotope, ^{14}C, but it is unstable and decays to ^{14}N.) Proportions of the two isotopes in a sample are customarily reported by an index, $\delta^{13}C$, which is calculated as follows:

$$\delta^{13}C = [(^{13}C/^{12}C)_{sample}/(^{13}C/^{12}C)_{standard} - 1] \times 10^3$$

Values are reported as per mille (‰) relative to a standard. For many biostratigraphic studies, the standards taken are the isotopic ratios of a fossil belemnite from the Cretaceous Pee Dee Formation of the Carolinas, or simply Pee Dee Belemnite (‰PBD). For other isotopes and studies, the standard might be the Standard Mean Oceanic Waters (‰SMOW), and so on.

Positive values of the index indicate an enrichment of the sample in the heavier isotope, ^{13}C, due to depletion of the lighter isotope, ^{12}C. Photoautotrophic organisms prefer the lighter isotope in photosynthesis; thus the enrichment of ^{13}C in a marine environment might indicate the activity of the phytoplankton, which is actively removing the lighter isotope. The decay of organic matter releases the lighter isotope into the environment. If this decay does not occur, as when large amounts of organic matter are deposited in sedimentary environments and buried before decay can release the organically fixed carbon, positive values of $\delta^{13}C$ are also produced. Negative values of the index indicate the opposite of the preceding situations. The heavier isotope in the sample is depleted, or diluted, because of the enrichment in the lighter isotope.

This brings us to the infamous "Strangelove Ocean" of Hsü and colleagues (1985). In modern oceans, a depth gradient exists in $\delta^{13}C$ values. Surface waters have $+\delta^{13}C$ values (because of photosynthetic activity of the phytoplankton) relative to deep waters, where the isotopic proportions between lighter and heavier carbon have a value of around zero in the $\delta^{13}C$ notation (figure 8.6). If the phytoplankton were to die (and thus cease ^{12}C uptake in surface waters), this depth gradient would collapse, and the resultant $\delta^{13}C$ values would be essentially flat from surface to depth (figure 8.6). This type of dead ocean, with no (or very little) phytoplankton activity, was named "Strangelove" by Hsü and colleagues (1985). The collapse of phytoplankton activity in the oceans would clearly indicate the occurrence of some catastrophic event, such as the impact of a bolide.

Expanding upon this model, Hsü and McKenzie (1990) propose the possibility of a postcatastrophic "Respiring Ocean," with a depth gradient in $\delta^{13}C$ that is the reverse of that found in modern living oceans (figure 8.6). In such an ocean, respiration processes are controlling surface-water $\delta^{13}C$ values, rather than photosynthetic pro-

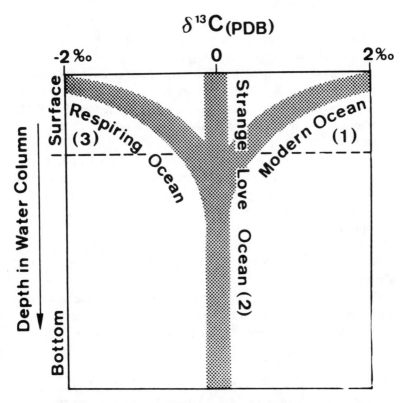

Figure 8.6. Depth variation of $\delta^{13}C$ values in a modern living ocean (1), a dead "Strange Love" ocean (2), and a postcatastrophic "respiring" ocean (3), as modeled by Hsü and McKenzie (1990). See text for discussion. (From Hsü and McKenzie 1990. Used with permission of the authors.)

cesses. In other words, dead phytoplankton are being decayed by bacterial activity, and $\delta^{13}C$ values are driven negative by the influx of ^{12}C produced by the microbial degradation of phytoplankton tissue. In summary, sudden negative shifts in $\delta^{13}C$ at biostratigraphic crisis horizons are potential signals of catastrophe, and thus they are of considerable interest in the search for evidence of impact.

Did a Strangelove Ocean, a dead ocean, exist at the end of the Frasnian? What was the state of the phytoplankton across the Frasnian/Famennian boundary? Carbon isotope anomalies will figure prominently in the next chapter, where the Frasnian/Famennian search turns out to be much more complicated (and interesting) than anyone could have predicted at the beginning.

The impact hypothesis predicts a period of global refrigeration following the vaporization of a large bolide and of the Earth at the target site. Which brings us to oxygen isotopes, because they are commonly used in trying to decipher temperature in the fossil record. Ratios of the isotopes ^{16}O and ^{18}O are customarily measured using the index discussed earlier for carbon isotopes and reported as $\delta^{18}O$. (Another isotope of oxygen, ^{17}O, exists, but it is very rare.)

Organisms that secrete calcium carbonate prefer the lighter isotope of oxygen, ^{16}O, in the biomineralization of their exoskeletons, that is, in the growth of the shell of a marine snail or clam. Thus they tend to remove the lighter isotope, leaving the heavier ^{18}O behind in the seawater. The fractionation of oxygen isotopes also occurs by inorganic processes. When water freezes, the lighter $H_2^{16}O$ is preferentially frozen out in the resultant ice, leaving behind the heavier $H_2^{18}O$ in the marine waters. Even evaporation preferentially removes the lighter $H_2^{16}O$.

The degree of fractionation of these isotopes in organic and inorganic carbonates is temperature sensitive (see Anderson and Arthur 1983; Anderson 1990) and generally follows the function:

$$T°C = 16.0 - 4.14\Delta + 0.13\Delta^2$$

where $\Delta = \delta^{18}O$ calcite (‰PDB) $- \delta^{18}O$ water (‰SMOW).

In general, positive values of $\delta^{18}O$ are indicative of lower temperatures, and negative values are indicative of higher temperatures. You can actually determine the temperature of the ocean millions of years ago, using the preceding equation, from a fossil shell composed of calcium carbonate—if you are very careful. Additional data must be present concerning the physiology of the organism that grew the shell and the value of $\delta^{18}O$ in the seawater in which the organism lived.

By far the worst problem with oxygen isotopic ratios is their notorious susceptibility to diagenetic alteration (as previously discussed in chapter 7). The analyst has to be very careful to ensure that the original $\delta^{18}O$ signature is preserved in the sample; otherwise the value found is meaningless and so is any temperature determination based upon it.

We have previously encountered oxygen isotopes and the Frasnian/ Famennian mass extinction in chapter 7, where they have been used to argue that the Earth became too hot (and not too cold, as predicted by the impact hypothesis) during the Frasnian/Famennian crisis interval. They will appear again in chapter 9.

Last, we discuss the isotopes of sulfur, ^{32}S and ^{34}S (sulfur actually has two other stable isotopes, ^{33}S and ^{36}S, but they are rare). The ratio $^{34}S:^{32}S$ is expressed with the same δ-parameter used for carbon and oxygen (see the preceding equation for carbon), and the standard used is that of the sulfur in the mineral triolite of the Canyon Diablo iron meteorite; thus values of $\delta^{34}S$ are reported (‰CDT). Sulfate-reducing bacteria prefer lighter isotopes, fractionating sulfur by splitting oxygen from sulfate ions, producing lighter isotope sulfides ($H_2{}^{32}S$) and residual heavy isotope sulfates (Faure 1986). High sulfide production or high pyrite burial will enrich seawater in the heavier isotope, ^{34}S, and yield positive values in $\delta^{34}S$. Positive $\delta^{34}S$ shifts in a basin may indicate the onset of stagnation, high evaporation, and warm-water conditions. High $\delta^{34}S$ values in surface waters may also indicate the upwelling of deeper-water anoxic layers, which are often poisonous.

Were the world's oceans stagnant and deadly during the Frasnian/ Famennian crisis? Or was there too much oceanic circulation, entire oceanic overturns in which poisonous bottom waters flooded the shallow marine shelves? And what, if anything, does any of this have to do with an asteroid impact? These questions will be explored in the next chapter.

In summary, isotopic ratios of carbon, oxygen, and sulfur may yield valuable clues to the biological and environmental conditions of the ancient Late Devonian oceans—their productivity, temperature, and degree of circulation. We shall encounter them all in the search for evidence of impact in chapter 9.

Tektites: Indisputable Evidence of Impact?

Tektites are silicate glass spherules produced by melting of the target site on the Earth by meteorite impact. Molten rock from the impact site is blasted through the atmosphere, cooling so rapidly that the melt forms glass and is shaped into characteristic sphere, pear, teardrop, and dumbbell shapes by their aerodynamic trajectories. Microscopically small tektites, generally less than 1 mm in diameter, are termed *microtektites*.

Tektites are well known from Cenozoic age strata. They are found spread over wide areas of the Earth's surface in fallout zones known as strewn fields (Glass 1982, 1990). The four known Cenozoic tektite strewn fields, and their ages, are the Australasian (0.7 Ma BP), Ivory Coast (ca. 1 Ma BP), Czechoslovakian (14.7 Ma BP), and North American (34 Ma BP). The "Czechoslovakian" strewn field is the result of fallout from the Ries impact crater in southern Germany, discussed in the preface of this book. Interestingly, known tektite fallout zones are not generally associated with iridium anomalies (Glass 1982, 1990), indicating that the Ries impact body, for example, was different in chemical composition from the bolide that collided with the Earth at the Cretaceous/Tertiary boundary.

Tektites may yield additional important evidence in addition to being indicators of an impact. Careful geographic mapping of variations in density of tektites found in standard units of sample, or variations in stratigraphic thickness of beds containing the tektites or other impact debris, may allow one to pinpoint the impact crater producing the tektites (or, if the crater has been eroded away, to state where it probably was located in the past). If not geochemically altered, composition of the glass itself may reveal information about the composition of the target rock, thus also assisting in tracking down the crater. Again, if the glass has not been geochemically altered, $^{40}Ar/^{39}Ar$ dating may reveal the geologic age of the impact event.

Glass is easily devitrified: chemically degraded, altered, and entirely leached away. For many years it was thought that the Cenozoic impact melt glasses were preserved chiefly because they were geologically still young. Tektites are now known from the Cretaceous/Tertiary boundary (Izett 1991; Smit et al. 1992), and we shall see in chapter 9 that preserved microtektites as old as 360 to 370 Ma are now being reported.

Care must be taken to determine whether any peculiar microspherules found in a sample are indeed microtektites. Morphologic comparisons alone are not sufficient proof, though they are encouraging if the morphologies found are of the "splash" form (teardrops, dumbbells, etc.). A recent study of multiple samples of Late Devonian sediments found four potential sources of microspherules: inorganically produced microspherules such as microtektites and cosmic spherules, organically produced microspherules such as conodont pearls, and microspheres produced by common laboratory contaminants (Wang and Chatterton 1993). Obviously, laboratory contaminants should be avoided from the onset. Organic structures such as conodont pearls are commonly flattened spheres, often with a central "dimple" depression, show concentric growth layering in thin section, and are composed of organic apatite compounds such as $Ca_5(PO_4)_3F$. Cosmic spherules are formed by the ablation of interplanetary dust and debris falling through the Earth's atmosphere and have ablation surface textures. They are either metallic or stony and are composed chiefly of iron compounds. Microtektites have "splash-form" morphology, often have spherical bubble cavities and lechatelierite inclusions in thin section, and are composed of acidic silicate glass (see Wang and Chatterton 1993).

To this list must be added another potential source of microspherules: volcanic glass. Volcanoes also produce melt rock glass, and, if the volcanism is explosive, they may produce glass bombs and droplets that are ejected into the atmosphere. Volcanic glass is water and volatile rich and usually contains phenocrysts and microliths of high-temperature minerals. Tektite glass is water and volatile poor, contains no phenocrysts or microliths, but may contain lechatelierite (quartz glass) and partially melted quartz grains (Glass 1990).

Impact craters are known from the Late Devonian, as discussed earlier in this chapter. Where there are impact craters, there should be impact melt glass. Do tektite layers exist in the strata deposited during the Frasnian/Famennian crisis interval? We shall see in chapter 9.

SUMMARY: SMOKING GUNS

The smoking-gun evidence is there. The Late Devonian is a time of at least three known continental impacts, probably as many as six, and who knows how many more oceanic impacts of which no evidence remains. The biological evidence is suggestive. There is the ammonoid Frasnes event in the early Frasnian, and there is the Flynn Creek impact crater (compare tables 3.1 and 8.4). There is the Kellwasser event at the end of the Frasnian and the Siljan impact crater. And there is the Hangenberg event at the end of the Famennian and the Charlevoix impact crater. The coincidence of these three craters occurring near the Givetian/Frasnian, Frasnian/Famennian, and Famennian/Tournaisian Stage boundaries is interesting (McGhee 1994). Are the stage boundaries themselves impact events?

More specifically, did *an impact* trigger the Frasnian/Famennian mass extinction? Probably not, unless the oblique impact model of Schultz and Gault (1990), with its extended-duration global cooling prediction, turns out to apply.

Did *several impacts* trigger the Frasnian/Famennian mass extinction? That scenario would indeed match the biological evidence. The surprisingly complex story of the search for the physical evidence of impact is taken up in the next chapter.

IX

The Search for Evidence of Impact

NEW YORK: THE BEGINNING

I first met meteoriticist Ed Olsen in 1981 while on research leave from Rutgers at the Field Museum of Natural History in Chicago. Although I traveled to Chicago to work with the museum's paleontologists, I actually spent more time discussing meteorite impacts with Ed Olsen. We eventually put together a research program that spanned the better part of the 1980s and had great fun doing so.

The debate concerning the Alvarez and colleagues (1980) paper was raging in the scientific community, and also in the popular press. At that time the debate centered around the true significance and origin of the "iridium anomaly" found by the Alvarez team at the Cretaceous/Tertiary boundary. Pro-impact and anti-impact interpretive papers rapidly emerged from the laboratories of scientists across the nation, and geologists around the world were traveling to their closest Cretaceous/Tertiary stratigraphic sections to see if the "irid-ium anomaly" was present there as well.

One day the obvious question arose in our discussions: Suppose there is an anomalous concentration of iridium in Frasnian/Famen-nian boundary sections as well? The potential significance of such a find was apparent, for it would not only lend powerful support to the original impact/extinction hypothesis of McLaren (1970), but would add fuel to the fire of debate concerning the significance of the "irid-ium anomaly" at the Cretaceous/Tertiary boundary itself.

The next obvious question was raised by Ed: Are there any com-plete Frasnian/Famennian boundary sections nearby? As a Devonian paleontologist, I knew the closest were in western New York State, and said so. That was not exactly "nearby" to Chicago, but when Dave Raup (who was then also at Field Museum) learned of our ideas, he strongly urged us to pursue them at once. Thus in the cold spring of 1981 we set out for New York State.

In 1981 the precise boundaries for the Frasnian and Famennian stages had still not been determined by international agreement. Three possible zonal candidates were being considered for marking the base of the Famennian (and hence the Frasnian/Famennian boundary), using the then current zonation: base of the Uppermost *gigas* Zone (= *linguiformis* Zone), base of the Lower (= Early) *triangularis* Zone, or base of the Middle *triangularis* Zone (see McLaren 1982).

The question of formal placement of the boundary relative to the conodont zonation was moot in western New York State, as cono-dont control was at that time virtually absent. What zonal control was present suggested that conodonts of the Late *triangularis* Zone occurred in the upper parts of the Dunkirk Black Shale in western New York State; thus these strata were clearly Famennian in age (Klapper et al. 1971). Conversely, ammonoids of the *Manticoceras* Stufe (of the older ammonoid Devonian zonation) occurred in the middle parts of the underlying Hanover Gray Shale; thus these strata were clearly Frasnian in age (House 1973).

Rather than worry about which conodont zone might fall where in the New York State stratigraphic section, given that the formal Fras-nian/Famennian boundary itself had not been internationally agreed upon, I decided to look for the most marked biotic turnover. The fact that a major drop in species diversity, and a marked change in faunal composition, occurs during the Late Devonian was recognized by paleontologists and stratigraphers in New York State well over a century ago (chiefly the work of the famous New York geologists James Hall and J. M. Clark in the middle 1800s; see also Chadwick 1935). This faunal break was utilized to divide the New York Late Devonian strata into two "series," now designated the Senecan and the Chatauquan, in New York terminology (Rickard 1975). Stratigraphically, this faunal break correlates with the basal Dunkirk Black Shale horizon in New York State (= "Chatauquan Series"). Strata below the Dunkirk Shale horizon (= "Senecan Series") contain an abundant and characteristic assemblage of Frasnian shellfish; these are absent and only a very depauperate fauna is preserved in strata immediately overlying the basal Dunkirk (Chadwick 1935; Dutro 1981). The existent ammonoid and conodont data (see above) demonstrated the approximate time equivalency of the Senecan/Chatauquan boundary in New York State with the Frasnian/Famennian boundary in Europe (if the faunal break was global and instantaneous, the two boundaries would be *precisely* the same).

Ed Olsen and I collected at two field sites that early spring and my wife Marae and I at another later in the same season, on a swing trip through New York between Chicago and our home in New Jersey. Ed Olsen had earlier met the geochemist R. Ganapathy while Ganapathy was doing research in Chicago. Ganapathy agreed to join our project and to perform the laboratory work in analyzing the Devonian samples for iridium concentrations. Disaster then struck in the form of a major labor dispute at the laboratories of J. T. Baker Chemical Company, where Ganapathy was based. The samples sat, unanalyzed, for months.

All three of us were at "Snowbird I" in the autumn, where I presented a brief paper outlining the project (McGhee et al. 1981). At the conference Ed and I met Carl Orth, who expressed interest in the project and offered to do the analytical work. Ganapathy agreed, and the samples were shipped to Orth at the Los Alamos National Laboratory.

At long last, the results began to come in. They were very disappointing. In not a single sample was any iridium detectable in the nanogram per gram concentration range. In contrast, iridium anomalies of 10–87 ng/g above background were then being reported at the Cretaceous/Tertiary boundary (Alvarez et al. 1982).

All the samples had been analyzed by instrumental neutron activation (INAA), which had a sensitivity of about 1 ng/g for iridium. Orth then processed a selected subset of the samples by radiochemical separations (RNAA), which were much more laborious and expensive, but had a sensitivity for iridium in the picogram per gram range.

The mean concentration of iridium found in the New York samples was only around 50 pg/g. The highest values found above this background concentration were 119 pg/g in a Dunkirk Beach sample, and 112 pg/g in a Walnut Creek Gorge sample (see figure 9.1). Concentrations of another siderophile element, cobalt, show only weak correlations with iridium (figure 9.1). In the deeper-water Dunkirk Beach and Walnut Creek Gorge sections cobalt appears to rise 30–40 μg/g above a background of 15–20 μg/g near our best placement of the Frasnian/Famennian boundary, but even that relationship is absent in the shallower water Mills section (figure 9.1).

We discussed these findings at the annual meeting of the Geological Society (Olsen et al. 1982) but delayed publishing the detailed results

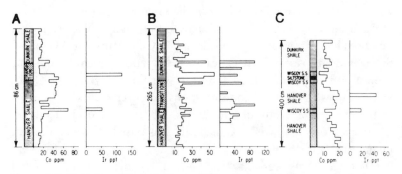

Figure 9.1. Stratigraphic and geographic variation of iridium and cobalt concentrations in the New York State sections. Deepest-water sediments are in the Dunkirk Beach (A) and Walnut Creek Gorge (B) sections in the western part of the state, where the Frasnian/Famennian boundary lies in the transition zone from the upper Hanover Gray Shale to the lower Dunkirk Black Shale. Shallower-water Wiscoy sands, which contain a Frasnian fauna in nearshore regions, occur in the Mills Mills (C) section to the east. Geographic distance from sections (A) to (C) is 102 km. (Modified from McGhee et al. 1984.)

at the moment. We had found no single "iridium spike," and our highest concentration was only 0.119 ng/g, a tiny fraction of values being reported at the Cretaceous/Tertiary boundary. But the possibility always existed that we had missed the "true" Frasnian/Famennian boundary (which at the time was still formally undefined), given the poor conodont zonal control in the New York State strata. Thirteen years after our initial study, this sad state of biostratigraphic affairs is finally being corrected. Jeff Over (1994) is conducting an extensive research program to locate the standard conodont zonal breaks in the New York State strata. Gratifyingly for us, Over (1994) has found that the Frasnian/Famennian boundary falls 10 cm below the base of the Dunkirk Black Shale at our Dunkirk Beach site, or exactly in our "transition zone" (figure 9.1).

The chapter is not yet closed on the New York story, however. Over and colleagues are also searching for iridium in sections additional to the three that we chose. To date, the highest value they have found is 0.38 ng/g, in a sample 17 cm above the Frasnian/Famennian boundary (Conaway et al. 1994).

In 1981 the search was for anomalous concentrations of iridium. Today, many other lines of physical evidence are being used in the search for impacts on the Earth in the geologic past (see chapter 8). No shocked minerals or microtektites have yet been found in New York (Katz and Over 1994), but the search has just begun.

BELGIUM: OFF TO EUROPE

In the summer of 1982 I planned to travel to Germany to work with my colleague Ulf Bayer at the University of Tübingen. We were interested in a group of Jurassic ammonites that persistently evolved similar morphologies in different lineages, an evolutionary phenomenon known as *iterative evolution*. I planned also to use the opportunity to continue the Frasnian/Famennian search.

The type area in Europe, where the terms *Frasnian* and *Famennian* were first defined, was in southern Belgium. Biostratigraphic control of the Belgian Frasnian/Famennian boundary was excellent, being tied to extinctions in both brachiopods and conodonts (Sartenaer 1968; Bouckaert et al. 1974).

Marae and I traveled up from Tübingen to visit the city of Brussels. There I met with Paul Sartenaer, director of the institute, and Pierre Bultynck at the Belgian Royal Institute of Natural Sciences, as previously arranged, to discuss which stratigraphic section would be best to sample. They decided on the Sinsin section, and Bultynck kindly drove down to southern Belgium with us to give us a description and tour of the outcrop. I sampled 4 m across the boundary, 20 cm up into the Famennian and 380 cm down into the Frasnian, and packed the samples off to Ed Olsen back in Chicago for preparation. They were then shipped to Carl Orth at Los Alamos for analysis.

The results were another disappointment (figure 9.2). There was no anomalous concentration of iridium in the section sampled. Interestingly, the background concentration of iridium was about the same as for New York, or 50 pg/g. The maximum value found for iridium was only 79 pg/g, however, compared to 119 pg/g in New York. Cobalt concentrations were essentially flat, with only one sample spiking at 60 µg/g above a background of about 20 µg/g. This peak did not correspond to the maximum iridium sample, which was above it stratigraphically (figure 9.2).

We had crossed the Frasnian/Famennian boundary four times, on two continents, and analyzed 410 samples. Given the biostratigraphic control on the Belgium section in particular, we were finally convinced there simply was no anomalous iridium concentration at the Frasnian/Famennian boundary. We wrote up the results and sent them on to *Nature*. From 1982 to 1988 other workers examined a few other Belgian sections in search of an iridium anomaly and also found nothing (Goodfellow et al. 1988).

The absence of an iridium anomaly does not *disprove* the impact of an extraterrestrial body; it simply does not provide any evidence *for* impact. The impacting bolide may have been a comet or a highly fractionated body, both being iridium poor. We specifically mentioned this possibility in conclusion, stating that we found no evidence of an impactor "*unless it were an ice-rich (stone-poor) comet*" (McGhee et al. 1984; italics mine). Almost exactly a decade later we were to see the real results of impact by a comet when the fragments of Shoemaker-Levy 9 exploded in the atmosphere of Jupiter in July of 1994. Back in 1984, we thought the Frasnian/Famennian story was over. As it turns out, it was just beginning.

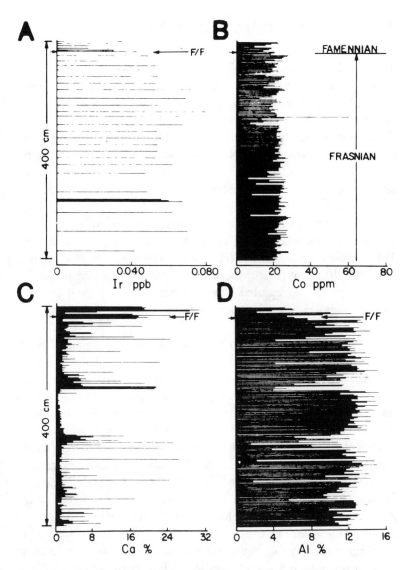

Figure 9.2. Stratigraphic variation of iridium, cobalt, calcium, and aluminum at the Sinsin section, southern Belgium. Position of the Frasnian/Famennian boundary after Sartenaer (1968) and Bouckaert et al. (1974). Calcium and aluminum vary sympathetically with carbonate/clay proportions in the section. (Modified from McGhee et al. 1984.)

AUSTRALIA: A SURPRISE FROM DOWN UNDER

The word was out: An iridium anomaly had been found at the Frasnian/Famennian boundary in Australia. At 300 pg/g it was a very small anomaly in absolute magnitude, a factor of 10 to 100 lower than the Cretaceous/Tertiary values, and only about 2.5 times higher than the maximum value found in New York. However, there was a big difference between the New York or Belgium sections and the Australian one. The excess iridium was concentrated in a single isolated peak that rose sharply above a flat background concentration of only 15 pg/g.

The word was also out that there were problems with the Australian iridium anomaly. The formal report appeared in late October 1984, in the pages of *Science* (Playford et al. 1984). The first problem was immediately apparent; the anomaly was in the wrong place! The iridium spike was reported from the basal beds of the Upper [= Late] *triangularis* Zone in a section in the Canning Basin in western Australia (figure 9.3).

In 1984 the formal placement of the Frasnian/Famennian boundary still had not occurred. Two horizons were under contentious discussion within the International Union of Geological Sciences, Subcommission on Devonian Stratigraphy: to place the formal base of the Famennian at the Middle *triangularis* Zone or to place it lower at the Early *triangularis* Zone. Those who favored the Middle *triangularis* Zone (hereafter called the "pragmatists") did so largely on the basis of practicality, as this zone is present in most stratigraphic sections around the world, whereas the Early *triangularis* Zone is missing in numerous sections. Those in favor of the Early *triangularis* Zone (hereafter called the "philosophers") were motivated more on philosophical than practical grounds. They wanted the placement of the stage boundary, an important marker in geologic time, to be as close as possible to a major event in the history of life on the Earth, namely, the mass extinction.

It was clear by 1984 that the main pulse of the extinction occurred near the Early *triangularis* Zone, perhaps taking place within the preceding *linguiformis* Zone (called the "Uppermost *gigas* Zone" at that time). McLaren (1982:477) himself concluded: "The horizon of

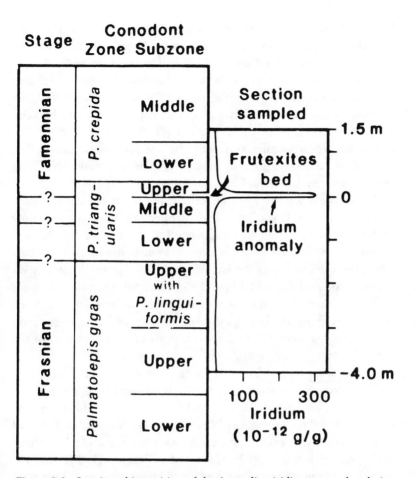

Figure 9.3. Stratigraphic position of the Australian iridium anomaly relative to the conodont zonation then in use (Ziegler 1971). Note also the position of the *Frutexites* bed, and uncertainties (question marks) at that time in the formal placement of the Frasnian/Famennian boundary. (From Playford et al. 1984. Copyright (c) 1984 by the American Association for the Advancement of Science. Used with permission.)

the event appears to lie most probably within the Uppermost *P. gigas* or Lower *Pa. triangularis* conodont subzone, but it might be as early as Upper *gigas* or as late as the base of the Middle *triangularis* subzones," a position he reiterated early in 1984 (McLaren 1984). Thus it was puzzling, to me at least, to see later in that same year

Digby McLaren as a coauthor on a paper which maintained that the extinction may have occurred as late as the Upper [= Late] *triangularis* Zone, where the iridium anomaly had been discovered. Early in the next year, in a sole-authored paper, McLaren (1985:171) wrote: "After initial failures (McGhee et al. 1984), the Canning Basin anomaly discussed in this paper was found by sampling at the horizon of the extinction (Playford et al., 1984). The above discussion justifies reviewing the possible contributions that accurate field observations in biostratigraphy may make to testing certain astrophysical hypotheses."

Those words were to prove prophetic, but not in the sense of their original intention. It was later demonstrated that the biostratigraphy of the *Australian* section was inaccurate and wrong, and that the beds containing the iridium anomaly occurred not in the Late *triangularis* Zone, but even higher in the section, at the level of the Early *crepida* Zone (Nicoll and Playford 1988, 1993). The final, accurate placement of the beds containing the anomaly indicates this horizon to be some three zones above the biological crisis horizon, and thus these strata were deposited some 1.5 to 2 million years after the horizon of the extinction (the *linguiformis* Zone).

The second problem had to do with the peculiar bed in which the iridium anomaly was found (figure 9.3). It is described as a prominent stromatolite bed, some 12 cm thick, composed of closely spaced microstromatolites of the fossil cyanobacterium *Frutexites* (Playford et al. 1984). It was known previously that these bacteria metabolized siderophile elements, and Playford and colleagues (1984) reported that the iridium was concentrated by a factor of 2 in filaments of *Frutexites* relative to the surrounding host rock.

The conclusion seemed to leap from the pages of the article: Iridium could be biologically concentrated! To understand the significance of the discovery of Playford and colleagues (1984), one has to think back to the intellectual climate of the early to middle 1980s. The pro-impact and anti-impact camps were in full battle, much of it centered at that time around the significance of iridium. The pro-impact side maintained that iridium, as a platinum group element, was geochemically inert. That is, it tended not to react with other elements, and hence had a very low mobility in stratigraphic sections. Therefore if one found large concentrations of the element at a given horizon, this meant that anomalous amounts of the element had been deposited at that horizon. A possible source for the deposition of anomalously large amounts of iridium was, of course, the vaporization of an iridium-rich bolide on impact with the Earth. The anti-impact side countered by proposing a series of hypothetical models to concentrate iridium at horizons by secondary geochemical or diagenetic processes, thus in effect denying that anomalous amounts of the element had been deposited primarily. If the concentrations were secondary and artifactual, then no bolide impact was needed to explain them.

The article of Playford and colleagues (1984) clearly showed that iridium abundances present in the ambient environment could be concentrated and enriched by biological processes. The fossil cyanobacterium *Frutexites* may have extracted the iridium, along with other siderophile elements, directly from normal seawater and concentrated these elements in its tissues. The high concentrations of iridium found in these rocks some 365 Ma later simply reflected the normal metabolic activity of these bacteria.

The possibility could not be ruled out, however, that iridium was present at higher than normal levels originally, and the bacteria simply enhanced this initial high concentration (Playford et al. 1984:439). Thus Playford and colleagues (1984) left open the possibility that the iridium anomaly could ultimately be traced back to an impact event. This possibility was expanded upon by McLaren, one of the coauthors of the paper, in a sole-authored paper published early in the next year (McLaren 1985).

In the summer of 1984 Carl Orth traveled with Phillip Playford to the Canning Basin, Australia, to seek and sample other exposures near the original McWhae Ridge site (Orth 1989; Orth et al. 1990). That same summer I was sampling an additional Frasnian/Famennian locality in Europe, this time in Germany (to be discussed below). In Australia, Orth and Playford found an exposure in a stream bed where *Frutexites* stromatolites alternated with other stromatolites over some 40 cm of section. Later analysis showed that the concentration of iridium and selected other elements varied directly in proportion to the density of *Frutexites* filaments (figure 9.4). Whenever a *Frutexites* bed appeared in the section, so did an iridium anomaly. "This observation, plus another case where the *Frutexites*/Ir correla-

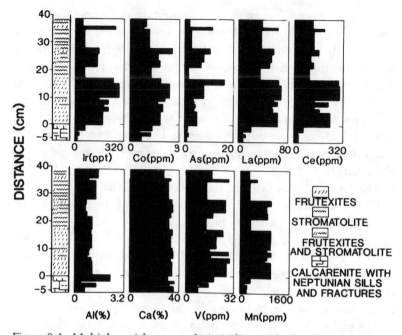

Figure 9.4. Multiple enrichment peaks in iridium and other elements in a creek bed section near the McWhae Ridge section, Australia. These enrichments clearly correlate with the presence of *Frutexites*, as other stromatolite beds without the bacterium are not enriched in these elements. (From Orth et al. 1990.)

tion occurred over a stratigraphic interval of over 1 m, provided evidence that the bacteria were merely concentrating the Ir and other elements out of seawater" (Orth 1990:48). These studies prompted one scientist to query "Iridium Anomalous No Longer?" in the pages of *Nature* (Donovan 1987), and sparked a series of papers examining the ability of both bacteria and fungi to concentrate iridium as well as the search for iridium anomalies in other fossil stromatolitic horizons not associated with mass extinctions (Dyer et al. 1989; Wallace et al. 1991).

A perennial question confronting every stratigrapher is the query, "How much time does this sedimentary unit represent?" In this case, how much time does 40 cm represent, particularly in condensed sections? Playford and colleagues (1984) suggested a minuscule rate of only 0.6 mm/ka for such condensed units. At that rate, the 40 cm would represent around 670 ka—that is, microbial iridium concentration for over half a million years in the Canning Basin. Geophysicists Hurley and Van der Voo (1990) became interested in this general question. They carefully sampled a 14.5 cm interval of the *Frutexites* stromatolites at the McWhae Ridge section for paleomagnetic analysis. The result of their study showed the existence of five separate magnetic polarity zones within this 14.5 cm interval, confirming that this small unit (only 14.5 cm thick) indeed represents a very long period of time (figure 9.5).

The story has become even more fascinating with the recent discovery of yet another iridium anomaly in the Canning Basin, this time in a section of drill core through these ancient sediments (Nicoll and Playford 1993). The new anomaly is Frasnian, but it is older than the biological crisis horizon and much older than the originally reported anomaly (figure 9.6). Most important, perhaps, is the fact that it also occurs in *Frutexites* microstromatolites (Nicoll and Playford 1993).

By the beginning of the 1990s it was clear to all that the original Australian iridium anomaly postdates the Frasnian/Famennian boundary and was formed by cyanobacterial concentration (Becker et al. 1991). Why was the original horizon even thought to be associated

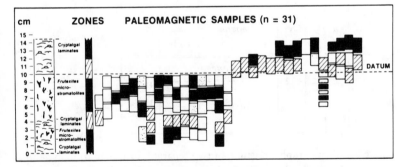

Figure 9.5. Stratigraphic distribution of five separate magnetic polarity zones within a 14.5 cm interval at the McWhae Ridge section, Australia. In the "Zones" column black represents normal polarity; diagonal lines represent reversed polarity. The upper *Frutexites* bed (at the datum line) is the horizon of the iridium anomaly reported by Playford et al. (1984). (From Hurley and Van der Voo 1990., Used with permission of the authors.)

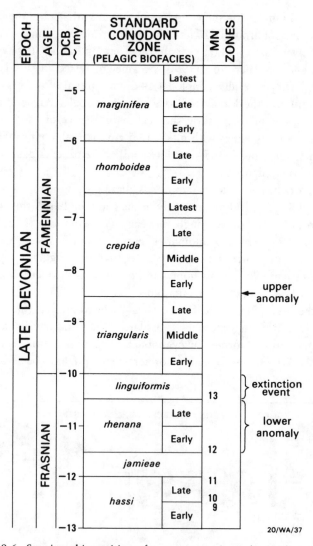

Figure 9.6. Stratigraphic position of two separate Australian iridium anomalies reported by Nicoll and Playford (1993). "MN Zones" gives the Montagne Noir conodont zonation of Klapper (1988), in addition to the standard. Both anomalies occur in *Frutexites* beds, and note that neither occurs at the extinction horizon (*linguiformis* Zone). The upper anomaly (Early *crepida* Zone) is the one originally reported by Playford et al. (1984) as occurring in the Late *triangularis* Zone. (From Nicoll and Playford 1993. Reproduced with the permission of the Executive Director, Australian Geological Survey Organization. Artwork courtesy of R. S. Nicoll.)

with an extinction? The answer: stable isotope anomalies. Playford and colleagues (1984) found independent isotopic evidence to suggest that something peculiar had indeed occurred at the horizon of the iridium anomaly. Both $\delta^{13}C$ and $\delta^{18}O$ show sharp decreases in values at the level of the originally reported iridium anomaly (figure 9.7). In

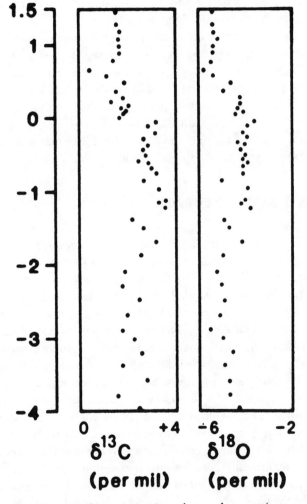

Figure 9.7. Stratigraphic variation in carbon and oxygen isotope ratios in the Australian McWhae Ridge section. Scale at left gives distance in meters above and below the iridium anomaly. Note the negative shift in $\delta^{13}C$ values at the level of the iridium anomaly. (Modified from Playford et al. 1984.)

particular, values of $\delta^{13}C$ undergo a -1.5 (‰PDB) shift in two pulses, one at the anomaly horizon and one about 75 cm above it, suggesting sharp declines or cessation of phytoplankton activity during this interval (a "strangelove ocean," see chapter 8). Values of $\delta^{18}O$ are more difficult to interpret (largely because of the problem of diagenetic overprint) but would be consistent with an increase in oceanic water temperatures at this horizon. Something seems to have happened in the Early *crepida* Zone, but what?

The search for impact evidence had expanded to include any peculiar microspheres that might be tektites, or impact-derived melt glass, by 1984. Playford and colleagues (1984) report of negative results in their search for microtektites at the horizon of the iridium anomaly (Early *crepida* Zone). Thus one might suppose the book finally to be closed on the possible early Famennian impact scenario. But it is not. Kun Wang was to find the microtektites in 1992. But that part of the story will have to wait for its proper time to be told.

GERMANY: BACK TO EUROPE

Word of the Australian iridium anomaly was widespread in early 1984, though the publication of the research was to appear only later in the year, in late October. In the spring of 1984 Jim Sorauf, a Devonian cnidarian specialist, and I convened a Paleontological Society symposium at Providence, Rhode Island, on the Frasnian/Famennian biotic crisis. Willi Ziegler, director of the Senckenberg Research Institute in Frankfurt, Germany, was at the meeting and we discussed the possibility of continuing the Frasnian/Famennian impact search in Germany. The European search was even more pressing now, given a positive find in Australia (or at least what at the time was considered to be a positive find) of an iridium anomaly precisely linked to a definite conodont horizon (that assessment was also to fall, as discussed earlier). He assured me that there was indeed an excellent section, stratigraphically complete, with conodont zonal control at the centimeter level. It was called Steinbruch Schmidt and was later argued to be the finest section in the world with regard to the designation of the formal boundary stratotype for the Frasnian/Famennian boundary.

In the summer of 1984 Marae and I traveled to the Steinbruch Schmidt site, near the city of Bad Wildungen, Germany. Willi Ziegler and Gil Klapper, an American conodont specialist who was working with Ziegler at the Senckenberg, accompanied us. Ziegler provided a detailed description and charts of the zonal breaks in the section and kindly marked off each conodont zone in outcrop. We all were particularly interested in two distinct horizons: the biological crisis horizon (the *linguiformis* Zone) and the horizon where the iridium anomaly had been (at that time) reported in Australia (the Late *triangularis* Zone).

I spent considerable time at the Steinbruch Schmidt site, sampling every single limestone bed and shale interval, beginning within the Late *triangularis* Zone and passing downward through the Late *rhenana* Zone—a continuous column of 5.01 m of rock. Back in Chicago, Ed Olsen painstakingly examined the samples for any signs of shocked minerals or other anomalous spherules and found nothing.

The geochemical results came from Carl Orth at Los Alamos: also nothing. There was no anomalous iridium in the section, not even at the Late *triangularis* Zone, where the Australian anomaly had been reported.

At first glance, fifteen peaks of iridium, from 75 to 159 pg/g, rise above a background level of 27 pg/g in the section (figure 9.8). This might be considered to be evidence of fifteen impact events spread over the duration of the crisis interval. Unfortunately, this is not the case. There is no anomalous iridium peak in the distribution, because the higher iridium values are clearly associated with the clay partings in the limestone sequence. Iridium correlates strongly with total aluminum from the clay minerals, so that a plot of the ratio of iridium to aluminum is virtually flat throughout the section (figure 9.8). Iridium plotted on a carbonate-free basis still reveals no anomalous peak (figure 9.8).

The crucial upper part of the section is given in figure 9.9. This part of the section contains both the "Upper Kellwasser Limestone" (more on this important unit follows later), which is the actual biological crisis horizon (within the *linguiformis* Zone) and also the horizon where the iridium anomaly was initially reported from Australia (top of the section). Iridium concentrations are flat at background levels at the top of the section, and the lower spikes in iridium concentrations

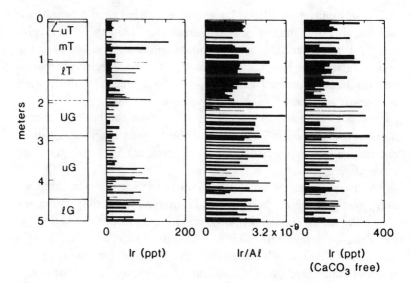

Figure 9.8. Stratigraphic variation in iridium and aluminum at the Steinbruch Schmidt section, Germany, relative to the conodont zonation then in use (Ziegler 1971). Zonal symbols: lG, uG, UG = Lower, Upper, Uppermost *gigas* Zone; lT, mT, uT = Lower, Middle, Upper *triangularis* Zone. Note that although fifteen peaks in iridium arise above background, the iridium is associated with clay partings in the limestone sequence such that the Ir/Al ratio is essentially flat. The Upper Kellwasser Limestone occurs within the interval of 1.4 m to 1.9 m (i.e., between solid line beneath lT to dashed line within UG). (Modified from McGhee et al. 1986a.)

are exactly matched with spikes in aluminum percentages, which in turn correspond to the presence of clay and shale partings in the overall limestone sequence (lithologies are given in the left part of figure 9.9).

At the time (in 1985) we were puzzled by the absence of an iridium anomaly in the top part of the section (Late *triangularis* Zone). The fallout from a global, impact-generated, dust cloud must have reached Europe, but where was it? We now know (as discussed earlier) that the iridium concentrations in Australia occur higher up in the section (in the Early *crepida* Zone); thus even if they were impact derived we would not have found them in this part of the section in Europe.

The section does contain stable isotope anomalies but not of the

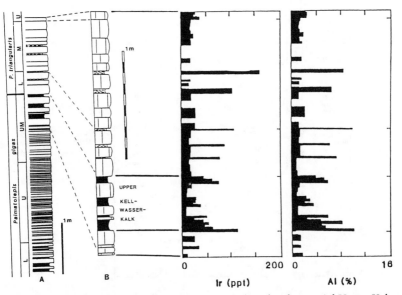

Figure 9.9. Iridium and aluminum abundance data for the crucial Upper Kellwasser Limestone (the extinction horizon, within the Uppermost *gigas* [= *linguiformis*] Zone) up to the originally reported level of the Australian iridium anomaly (Upper [= Late] *triangularis* Zone) of Playford et al. (1984). (Modified from McGhee et al. 1986b.)

type we expected. Quite the reverse, in fact. Values of $\delta^{13}C$ rise about +1 in the bottom part of the section, to 2.0 (‰PDB), decline to a low of 0.5 (‰PDB), then $\delta^{13}C$ sharply jumps +2.5 to values of 3.0 (‰PDB) at the biological crisis horizon (figure 9.10). This time they do not decline, but remain elevated throughout the remainder of the section sampled.

Values of $\delta^{18}O$ also show a positive shift of about +1 (figure 9.10). Although the latter pattern may be used to suggest a decline in oceanic water temperatures at the critical horizon (as suggested by the ecological data; see chapter 6), the susceptibility of $\delta^{18}O$ values to diagenetic alteration prompted us to avoid making too much of this +1 $\delta^{18}O$ shift (McGhee et al. 1986a:779). It is considered anomalous by Goodfellow and colleagues (1988).

Thus, rather than the expected sharp drop in $\delta^{13}C$ at the biological crisis horizon, which might have signaled the collapse of phytoplankton productivity ("a strangelove ocean;" see chapter 8), we see the

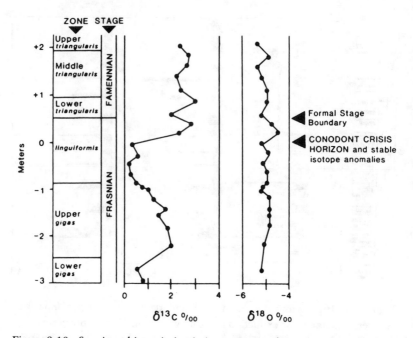

Figure 9.10. Stratigraphic variation in isotope ratios for carbon and oxygen at the Steinbruch Schmidt section. Note the sharp positive increase in $\delta^{13}C$ values at the base of the Upper Kellwasser Limestone horizon (the conodont crisis horizon, position 0 m in the *linguiformis* Zone), and the much smaller positive shift in $\delta^{18}O$ values at the same level. (Modified from McGhee 1989b.)

exact opposite. The sudden increase in $\delta^{13}C$ ratios at the critical horizon suggests that either a bloom in phytoplankton activity or upwelling of deeper oceanic waters, or both, occurred in this interval of time. Phytoplankton blooms and upwelling episodes are usually local or regional, and not simultaneously global, events in the present world. We noted that there had been only one other study, which suggested an almost identical signature of $\delta^{13}C$, in neighboring Belgium by Dunn and colleagues (1985). Thus we concluded that our unexpected sharp positive shifts in $\delta^{13}C$ could have been only local phenomena, not global, and that even though they do occur at the biological crisis horizon, they may not tell what triggered the global nature of this event (McGhee et al. 1986a,b).

Werner Buggisch (1972) has long been interested in the peculiar "Kellwasser" black shale and bituminous limestone horizons in the

European and Near Eastern late Frasnian strata and had undertaken a comprehensive geochemical study of the twin "Kellwasserkalke" a decade before (see chapter 7). Upon seeing the results from the Steinbruch Schmidt section he and his student Michael Joachimski began to examine systematically the stable isotope geochemistry of other Frasnian/Famennian sections (Joachimski and Buggisch 1992, 1993). They found that Steinbruch Schmidt was not unique. In additional sections in Germany, France, and Austria, an abrupt positive shift occurs in $\delta^{13}C$ at the biological crisis horizon, and these values remain elevated for at least three conodont zones into the Famennian (figure 9.11). Positive shifts in $\delta^{13}C$ have also now been reported from Frasnian/Famennian outcrop and borehole sections in Poland (Narkiewicz 1992; Halas et al. 1992). All across Europe, $\delta^{13}C$ values show a sharp positive excursion of up to +2.8 at the Frasnian/Famennian boundary, over a general Frasnian background value of 1.0 (‰PDB).

A section in the Carnic Alps, the Wolayer Gletscher, is of particular importance. The Kellwasser limestones were deposited under anoxic conditions, and their black shales and bituminous limestones indicate

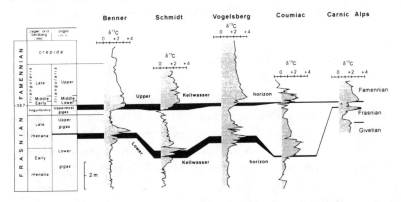

Figure 9.11. Geographic and stratigraphic variation in carbon isotope ratios in additional Frasnian/Famennian boundary sections in Germany, France, and Austria. The standard conodont zonation (Ziegler and Sandberg 1990) and the zonation in use in 1986 (Ziegler 1971; see table 1.7 of this book for zonal comparisons) are both given. Thus compare these 1993 data with those of 1986 given in the previous figure. (From Joachimski and Buggisch 1993. Artwork courtesy of M. M. Joachimski.)

the burial of large amounts of organic carbon. There are no Kellwasser horizons at the Wolayer Gletscher site. Positive excursions in $\delta^{13}C$ still occur at the equivalent time horizons of the two Kellwasser limestones (Late *rhenana* Zone and *linguiformis* Zone; see figure 9.11). The Wolayer Gletscher site proves that the positive anomalies in $\delta^{13}C$ were not simply developed under anoxic conditions. Thus Joachimski and Buggisch (1993) argue that the European positive excursions do reflect changes in the $\delta^{13}C$ of the total dissolved carbon reservoir of marine waters. Additional sections have now been reported from Australia and Nevada, USA, that also possess heavy carbon excursions without accompanying anoxic sediments (Joachimski and Buggisch 1994).

The question still remains: "Were these changes really global?" If so, positive excursions in $\delta^{13}C$ should occur at every Frasnian/Famennian boundary section around the world. To the contrary, negative excursions (Goodfellow et al. 1988) and no change at all (Geldsetzer et al. 1987) have been reported from sites in Canada. On the other side of the world, a sharp negative $\delta^{13}C$ shift at the Frasnian/Famennian boundary in China has been reported by Wang and colleagues (1991). Needless to say, the search still continues, and the mystery of the Frasnian/Famennian carbon isotope anomalies remains unsolved.

CANADA: THE SEARCH WIDENS

Up to 1985 the search for impact evidence had centered on the Late Devonian strata of Europe, eastern North America, and western Australia. The search was soon to expand to western North America and southern China. Thick sequences of Late Devonian sedimentary rocks are found in the Rocky Mountains of Canada and the United States. Unfortunately, many of those sequences had poor biostratigraphic control.

Biostratigraphic matters were not helped any by continued—and contentious—debate concerning the formal designation of the Frasnian/Famennian boundary in the years 1983 to 1987. Of the two proposals under consideration by the International Union of Geological Sciences, Subcommission on Devonian Stratigraphy (see the preceding discussion), the pragmatists won the first decision in 1983 and the subcommission recommended placing the base of the Famennian at the base of the Middle *triangularis* Zone (see Ziegler and Klapper 1985; House 1988b). Vehement protest arose immediately from the philosophical wing, and the debate resumed.

In the Canadian Rockies, the Frasnian/Famennian boundary appears to fall at the contact between the Ronde Formation and the overlying Sassenach Formation. Conodonts of the Upper *gigas* Zone [= Late *rhenana* Zone] are reported from the lower Ronde, and conodonts of the Upper [= Late] *triangularis* Zone from the upper Sassenach (Geldsetzer et al. 1987). The main question concerns whether any section is missing between these two zonal horizons, particularly the critical *linguiformis* Zone. A site at Medicine Lake, Alberta, was chosen by Geldsetzer and colleagues (1987) for geochemical analysis.

They found no iridium anomaly. Iridium concentrations fluctuate around a background of about 20 pg/g, with increases at the Mt. Hawk/Ronde and Ronde/Sassenach formational contacts (figure 9.12). The latter contact is considered the horizon of the extinction

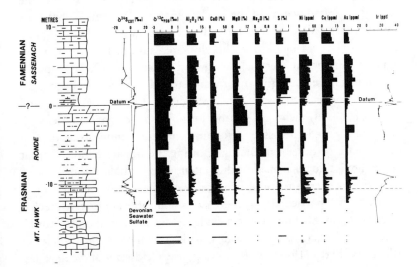

Figure 9.12. Stratigraphic variation in sulfur and carbon isotope ratios (left side of figure), selected elemental abundances, and iridium concentrations (right side of figure) in the Medicine Lake section, western Canada. (From Geldsetzer et al. 1987. Artwork courtesy of H. H. J. Geldsetzer.)

event; however, the increase in iridium (to about 30 to 35 pg/g) is not just at the contact, but remains high up into the Sassenach Formation. Carbon and oxygen isotopic analyses were also unfruitful. Unlike the European sections, $\delta^{13}C$ values remain flat across the Frasnian/Famennian boundary (figure 9.12), showing neither positive nor negative excursions. Analyses for $\delta^{18}O$ were conducted, but the resulting values varied erratically and were not interpretable (Geldsetzer et al. 1987:395).

A new isotopic analysis was introduced in the Canadian study, that for $\delta^{34}S$. Values of $\delta^{34}S$ begin negative, increase slightly upsection, but jump sharply to +20.8 (‰CDT) in a 5-cm-thick pyrite unit at the base of the Sassenach Formation (figure 9.12). Values remain positive for about 2 m above the contact, then turn negative again. Since their original study, sharp increases in $\delta^{34}S$ at the Frasnian/Famennian boundary have also been reported in outcrop and borehole sections in Poland (Narkiewicz 1992; Halas et al. 1992).

Geldsetzer and colleagues (1987) argue that an abrupt onset of anoxic conditions is evidenced by the sharp $\delta^{34}C$ spike, consistent with the sudden injection of deep anoxic bottom waters into the normally aerated waters of the shallow marine platform. Such waters are hypothesized to have poisoned the entire water column, killing both benthos and plankton. The ultimate cause of the anoxic pulse remained unknown, but Geldsetzer and colleagues (1987) proposed that it may have been triggered by an asteroid or cometary impact.

Their impact proposal drew immediate response by Wilde and Berry (1988:86), who called the proposal a "convenient deus ex machina for the Kellwasser Devonian extinction" and insisted that, in the absence of other evidence supporting the impact hypothesis, "it seems advisable to consider alternatives to the bolide impact as a causative mechanism for driving anoxic water onto the shelf on a global basis," particularly as there exist "plausible alternative mechanisms with supporting geologic evidence available in the literature." One such alternative mechanism was the oceanic overturn model of Wilde and Berry (1984, 1986). In response, Geldsetzer and colleagues (1988) question the applicability of this model, which they term the "climatically controlled cooling theory," in explaining the Frasnian/Famennian extinction, and conclude, "We continue to consider the impact theory, not as a convenient deus ex machina but as a solution with a higher degree of probability than the climatically controlled cooling theory."

CHINA: IRIDIUM AT LAST?

The debate over the formal placement of the Frasnian/Famennian boundary was at long last over. Perhaps. After much debate, the boundary-event philosophers emerged in the fore. The International Union of Geological Sciences, Subcommission on Devonian Stratigraphy, revoked its 1983 decision in favor of the Middle *triangularis* Zone, and voted to placed the base of the Famennian at the base of the Lower *triangularis* Zone (see House 1988b; Cowie et al. 1989; Oliver and Chlupac 1991). The now famous Steinbruch Schmidt site in Germany, which the reader has encountered many times thus far, lost out in the final vote and the GSSP ("Global Stratotype Section and Point") was placed in the Coumiac section in the southeastern Montagne Noire of southern France (Klapper et al. 1993).

Dissent abounded elsewhere, however. A new report of an iridium anomaly at the Frasnian/Famennian boundary was out, this time in China. The paper, which appeared in the pages of *Science*, was very unusual: The authors themselves could not agree on the conclusions and actually indicated the division of opinion among them, by name, at the end of the paper.

An iridium spike of 350 pg/g, against a background of 16 pg/g, had been found by Wang and colleagues (1991) in Xiangtian, Guangxi, southern China. One could now say that the anomaly was indeed at the Frasnian/Famennian boundary, both in an informal and formal sense, as the stage boundary was now placed at the *linguiformis*/ Lower *triangularis* zonal boundary, and at the horizon of the main pulse of the extinction.

The iridium, and many other elements as well, is enriched in a 20-cm-thick mudstone labeled "Bed E" (figure 9.13). This unit is barren of fossils and is argued by both Wang and colleagues (1991) and Yan and colleagues (1993) to have been deposited in a "Strangelove ocean" environment, virtually devoid of phytoplankton activity. The iridium anomaly is not a large one (230 pg/g; 350 pg/g only on a carbonate-free basis), but it stands out sharply when compared to the

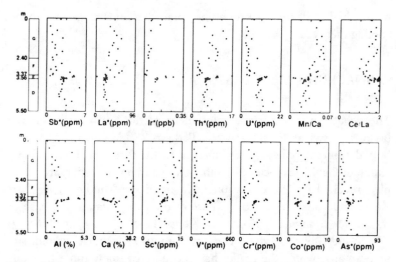

Figure 9.13. Stratigraphic variation in selected elemental abundances and ratios in the Xiangtian section, southern China. Iridium concentrations are given in the top row, third column from the left. Asterisks indicate abundances reported on a carbonate-free basis. Bed F marks the base of the Famennian, beds E and D comprise the *linguiformis* Zone (latest Frasnian). (From Wang et al. 1991. Used with permission of the author.)

low background concentrations of iridium. Values of $\delta^{13}C$ show sharp −3.49 negative excursions, from +1 (‰PDB) to −2.49 (‰PDB), at the same horizon as the iridium spike (figure 9.14). Values of $\delta^{18}O$ show slight negative shifts. Examination of samples for the presence of shocked minerals or microtektites turned up nothing (Wang et al. 1991).

Yan and colleagues (1993) conducted much more extensive isotopic analyses of the same section and obtained somewhat different results from those of Wang and colleagues (1991). They report a multipeak negative excursion of $\delta^{13}C$, with much larger negative values, up to −6.6 (‰PDB), all within "Bed E." Oxygen isotopic results are also different, in that they report a positive excursion of $\delta^{18}O$ of about +2, from around −6.5 (‰PBD) to −4.5 (‰PDB), within "Bed E" (figure 9.15). This positive shift in $\delta^{18}O$ is twice the magnitude we found at the same time horizon in Germany (McGhee et al. 1986a,b). Yan and colleagues (1993) consider the positive shift in $\delta^{18}O$ to indicate that a decrease in oceanic temperature of "several degrees centigrade" occurred during the deposition of "Bed E." These data Yan and colleagues (1993) found compatible with rapid, short-term global cooling and simultaneous decimation of the phytoplankton, both consistent with the expected effects of the impact of a large extraterrestrial body.

In conclusion, the majority opinion of the original 1991 *Science* paper was that the iridium anomaly at the Frasnian/Famennian boundary in southern China was real, and most probably due to bolide impact, similar to that at the Cretaceous/Tertiary boundary but smaller in magnitude (authors Wang, Chatterton, Hou, and Geldsetzer). The minority opinion noted the small magnitude of the iridium enrichment and consider that it and the other elemental patterns could be attributable to the reducing geochemical conditions present during deposition of the boundary sediments (authors Orth, Attrep).

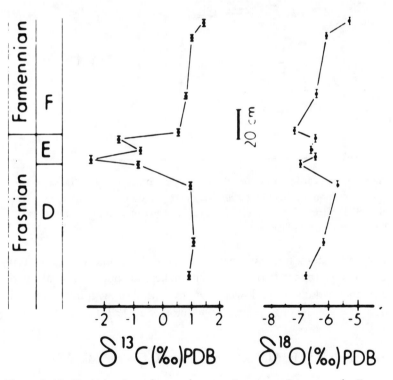

Figure 9.14. Variation in carbon and oxygen isotope ratios across the Frasnian/Famennian boundary, Xiangtian section. (From Wang et al. (1991). Used with permission of the author.)

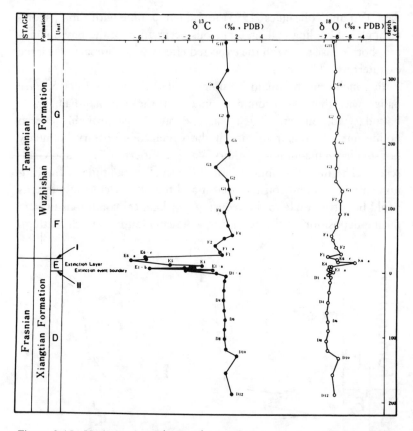

Figure 9.15. Variation in carbon and oxygen isotope ratios in the same Xiangtian section, but from later analyses performed in greater detail by Yan et al. (1993). From Yan et al. (1993). Used with permission of the authors and Elsevier Science Publishers.)

The story does not end there, however. Preliminary results from China now report *two* iridium anomalies, an upper anomaly with 226 pg/g iridium and a lower anomaly with 316 pg/g, both of which occur *within* the *linguiformis* Zone (Hou et al. 1992).

MULTIPLE SMOKING GUNS: IMPACT MELT GLASSES

It was the year 1992, and the word was out. The word, in this case, was *microtektites*. Word of the existence of impact melt glass at, or very near, the Frasnian/Famennian boundary in Belgium circulated rapidly at the fifth International Geological Correlation Program 216 Conference on Bio-Events in Göttingen, Germany, in early February of 1992. The Belgian Royal Institute of Natural Sciences' Paul Sartenaer, who had kindly assisted us in our initial search for iridium in Belgium back in 1982, was at the conference and described the discovery to me. A member of the Royal Institute, Jean-Georges Casier, had discovered what appeared to be microtektites in the Early *triangularis* Zone. At the same meeting, Kun Wang described to me microtektites he had also discovered in China—but in the Early *crepida* Zone!

A virtual explosion of new information in the search for impact evidence was to occur in 1992. In February of 1992, the discovery of the Chinese microtektites was announced at the International Geological Correlation Program 216 Conference on Bio-Events in Göttingen, Germany (Wang 1992a; Wang and Geldsetzer 1992). In May of 1992 Jean-Georges Casier and colleagues announced the discovery of the Belgian microtektites at a meeting of the American Geophysical Union in Montreal, Canada (Claeys et al. 1992a). At the same meeting, Boundy-Sanders (1992) announced the discovery of possible microtektites of Late Devonian age from western Canada. The latter report turned out to be a false alarm, as it was later discovered that the glassy spheres were recent surficial contaminants (Boundy-Sanders 1994). Formal reports and descriptions of the first two Late Devonian microtektite announcements appeared in papers published in *Science* later in 1992—the Chinese discovery on June 12th and the Belgian discovery on August 21st.

As the paper reporting the Chinese microtektites was published first, I will begin with the Chinese discovery, even though it is clear that these microtektites are considerably above, and postdate, the Frasnian/Famennian boundary. They do, however, reopen the Pandora's box of the original Australian iridium anomaly.

Wang (1992b) discovered microspherules in a marine limestone immediately below a 3-cm-thick claystone showing an anomalous enrichment in iridium and other elements in a section near Qidong, Hunan, in southern China. The iridium anomaly is very small, only some 40 pg/g against a background of around 15 pg/g, with a peak/background ratio of only 2.66 (figure 9.16). Equally small concentration shifts occur in other siderophile and chalcophile elements at the

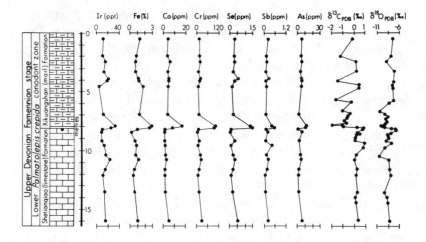

Figure 9.16. Stratigraphic variation in iridium (left side of figure) and other elemental abundances in the Qidong section, southern China. Note that the entire section is Famennian in age and that the anomalies occur in the Lower [= Early] *crepida* Zone, some four zones above the Frasnian/Famennian boundary. (From Wang et al. 1994.Artwork courtesy of K. Wang.)

same horizon, but none would have sparked comment were it not for the microspheres in the layer immediately underneath.

The microspheres show the typical "splash form" morphology of microtektites (figure 9.17). They are silicate glass, are volatile-poor, and contain lechatelierite, and thus are argued not to be of volcanic origin (Wang 1992b).

Most perplexingly, the microtektites occur well above the Frasnian/Famennian boundary, in the Early *crepida* Zone (Wang 1992b). The impact producing them would have occurred some 1.5–2.0 million years after the main biological crisis horizon in the older *linguiformis* Zone. Biologically, nothing much seems to have happened even in China at the time, for only a minor change in the brachiopod faunas is noted (Wang 1992b). The question is, "What else happened in this region of the world during the Early *crepida* Zone?" The answer: the Australian iridium anomaly!

Wang (1992b) has revived the hypothesis that the microbial concentration of iridium in the Australian anomaly occurred because there was excess iridium in the environment for the bacteria to concentrate. He proposes that a small bolide impact occurred in China during the Early *crepida* Zone, producing microtektite fallout in China and iridium fallout in neighboring Australia. Biological response to this impact may be reflected in the reported changeover in

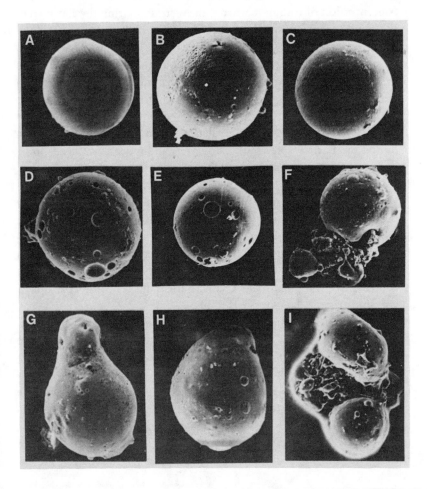

Figure 9.17. Scanning electron micrographs of the Qidong microtektites. Typical forms are spherical, tear-drop, or pear-shaped, or compound. (From Wang et al. 1994. Photograph courtesy of K. Wang.)

brachiopod faunas in China and the negative $\delta^{13}C$ excursion in Australia. He has even identified a likely candidate for the resultant impact crater: Taihu Lake, a 70 km structure of probable impact origin and of likely Late Devonian age, located about 900 km to the northeast of the Qidong site (Wang 1992a).

Substantial problems remain with this scenario, however, not the least of which is the long duration of time over which the microbial concentration of iridium occurred in Australia during the Early *crepida* Zone. Could a single impact in China have produced iridium fallout for more than half a million years in Australia? There is also now the second Australian iridium anomaly, in the Frasnian Late *rhenana* Zone. If the ancient cyanobacteria only concentrated iridium when excess iridium was present in the environment, then a possible impact source for this anomaly is also called for. The microtektites themselves seem unequivocal evidence for an impact event in China during the Early *crepida* Zone, however, and more work needs to be done in obtaining a precise age for the Taihu Lake structure.

The Belgian microtektites were actually first discovered by Jean-Georges Casier back in 1987, but he was informed by a colleague that they were volcanic in origin, and therefore nothing unusual (see report in Kerr 1992). They are of the typical "splash form" morphology of microtektites (figure 9.18), and subsequent chemical analysis reveals them to be silica-rich glass with very low water and volatile content, with no phenocysts or microliths, and thus arguably not of volcanic origin (Claeys et al. 1992b).

The microtektites occur at the Senzeilles section in a 5 to 10 cm layer located about 7 m above the transition between the Matagne and Senzeilles shales (figure 9.19). The Senzeilles Shale has been determined to occur in the Early *triangularis* Zone, and the Frasnian/Famennian boundary has traditionally been placed at the base of the transition shale zone (figure 9.19). Finer-scale stratigraphic control is not available at present, as conodonts are scarce in the section (Kerr 1992). The 7 m interval of sediment could represent several hundred thousand years, depending upon the rate at which the shales were deposited.

The microtektites are abundant in the spherule layer, and Claeys and colleagues (1992b) calculate an average concentration of 0.03 mg of glass per gram of sediment at this horizon. As the microtektite

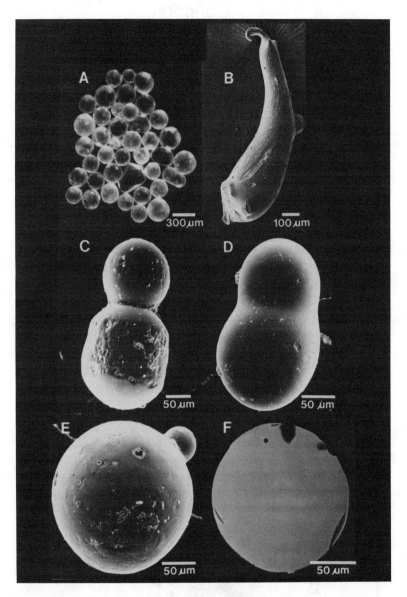

Figure 9.18. Light microscope photograph (A) and scanning electron micrographs (B-F) of the Senzeilles microtektites. (From Claeys et al. 1992b. Copyright (c) 1992 by the American Association for the Advancement of Science. Used with permission.)

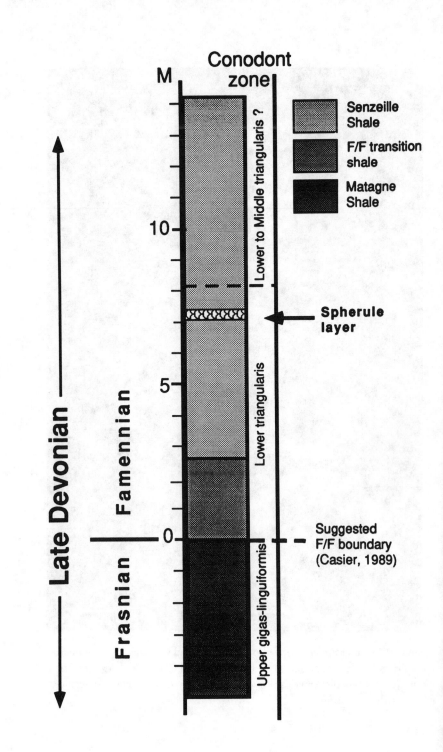

layer occurs very near, if not at, the Frasnian/Famennian boundary of 367.0 Ma BP (Harland et al. 1989), the obvious candidate for the source of the impact melt glass is the Siljan Crater, dated at 368 ± 1.0 Ma BP (Grieve and Robertson 1987), located just to the north of Belgium in Sweden (see chapter 8).

Later in 1992 the microtektite layer was found at a new site in Belgium, the Hony section (Claeys et al. 1994, Claeys and Casier 1994). A second find of preserved microtektites of such ancient age is in and of itself important, but there was more. This time the microtektite horizon appears to be only 1 m above the traditional placement of the Frasnian/Famennian boundary, prompting the *Science* declaration "Ancient Impact-Extinction Link Tightened in Belgium" (see report of Kerr 1993a).

Where is the precise Frasnian/Famennian boundary, and where does the microtektite layer occur relative to it? In neither Belgian section have conodonts been found in the critical shales just above and below the microtektite layer; thus precise zonal placement of the horizon remains frustrated at present (Claeys et al. 1992b, 1994). The microtektites at the two sites are similar in composition (Claeys et al. 1994; Claeys and Casier 1994); thus it appears the spherule layer is the same time horizon in the two sections. The apparent 7 m separation of the microtektite layer above the traditional Frasnian/Famennian boundary at the Senzeilles section and 1 m separation at the Hony section may simply reflect differing rates of sedimentation at the two sites. Alternatively, the traditional placement of the Frasnian/Famennian boundary at the two sites may be slightly off, particularly as conodont control is not present in microstratigraphic detail.

Claeys and Casier (1994) and Claeys and colleagues (1994) adopt the latter point of view, noting that Sandberg and colleagues (1988:282) suggested a placement of the base of the Lower [= Early] *triangularis* Zone at the Hony railroad cut only 2–3 cm below the bed in which the microtektites were later found. Thus the 1 m separation of the microtektite layer above the Frasnian/Famennian bound-

Figure 9.19. Stratigraphic position of the microtektites found in the Senzeilles section, southern Belgium. Note that the microtektites are earliest Famennian in age (Lower [= Early] *triangularis* Zone), possibly 7 m above the Frasnian/Famennian boundary in this section. (Modified from Claeys et al. 1992b. Artwork courtesy of P. Claeys.)

ary (labeled "FFB" in figure 9.20) may now have been reduced to 2 or 3 cm at the Hony section.

Geochemical analyses at the Hony section suggest a possible iridium anomaly may also be present at the microtektite horizon (figure 9.20). Claeys and colleagues (1994) report a spike in iridium concentrations of 83 pg/g, against a background of 40 pg/g, at the microtektite horizon ("FFB" in figure 9.20). However, two spikes of similar magnitude occur above the spherule layer and are not associated with microtektites (see figure 9.20). The small magnitude of the peak, a peak/background ratio of only 2.07, and the presence of two other peaks renders the significance of the iridium profile inconclusive. In fact, subsequent geochemical analyses by Herbosch and colleagues (1994) conclude there is "no Ir anomaly associated with the microtektite-like glass" in the Hony section.

Back in 1982 I had the suspicion that the most severe pulse of extinction during the Late Devonian came somewhat before the horizon of the base of the Famennian. Subsequent resolution of the biological data has proven that suspicion to be correct (see chapter 3). If the most severe pulse of extinction was impact triggered, it is there that one should search for impact evidence. With this idea in mind I sampled only 20 cm up into the Famennian at our Sinsin site in Belgium, but down for 380 cm into the Frasnian. We found no anomalous iridium, as described earlier.

To date no one has reported finding microtektites at the Sinsin section (although the search is actively on, as communicated to me by Paul Sartenaer at a meeting we attended in Plymouth, England, in September 1994). I still wonder, after all these years, whether the original story would have turned out the same if I had reversed the sampling scheme and sampled extensively up from the boundary into the Famennian rather than down from the boundary into the Frasnian, back in 1982. We may soon have the answer.

FRANCE: ALL QUIET AT THE GLOBAL BOUNDARY STRATOTYPE?

The cold fact remains, however, that both the Belgian and Chinese microtektite layers discussed earlier occur in stratigraphic horizons

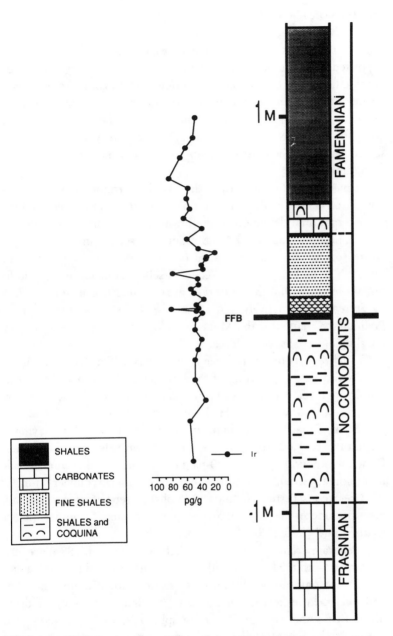

Figure 9.20. Iridium abundances (note the scale is reversed, increases are to the left) 1 m above and below the possible Frasnian/Famennian boundary ("FFB") in the Hony section, southern Belgium. Microtektites are found in the bed immediately above the datum marked FFB. (From Claeys et al. 1994. Artwork courtesy of P. Claeys.)

(Early *triangularis* and Early *crepida* zones, respectively) that were deposited *after* the deposition of the maximum diversity loss stratigraphic horizons (Late *rhenana* and *linguiformis* zones, respectively) during the Frasnian/Famennian crisis. They are both early Famennian in age, and although they both occur during the Frasnian/Famennian crisis interval (see tables 1.7 and 3.2), they are younger than the severest pulse of the mass extinction, which occurs in the latest Frasnian (table 3.2).

As mentioned briefly earlier, the "Global Stratotype Section and Point" (or GSSPs, as they are informally referred to by geologists) for the international definition of the Frasnian/Famennian Stage boundary has been formally placed in the Coumiac section in the Montagne Noire region of southern France (see discussion in Klapper et al. 1993). The global stratotype section itself (Coumiac) and an additional section within the stratotype area of the Montagne Noire, the La Serre site, are actively under analysis as this book goes to press. Preliminary results for the Montagne Noire sites are at present very puzzling indeed, as we shall see.

Verbal reports of a very large iridium anomaly at La Serre spread rapidly through the meeting of the International Geological Correlation Program 216 held at Calgary, Canada, in 1991. Discovered by W. D. Goodfellow and H. H. J. Geldsetzer, both of the Geological Survey of Canada, the iridium anomaly was reported as having a magnitude of 5,600 pg/g in a 4 cm layer directly at the Frasnian/Famennian boundary. At 5,600 pg/g this ancient Late Devonian iridium anomaly would rival those found in the much younger strata at the Cretaceous/Tertiary boundary! Published reports of the anomaly first appeared in the pages of *Science* with the query "Another Impact Extinction?" (Kerr 1992).

A separate French team of researchers (Girard et al. 1993) are also at work at both the La Serre and Coumiac sites. Girard and colleagues (1993) not only are analyzing these sections for iridium anomalies and microtektite layers, but are also searching for highly oxidized nickel-rich spinels or spinel-bearing spherules, which are now also thought to be another indicator of cosmic, or extraterrestrial, material in Earth sediments.

To date their search has revealed no evidence of impact in the latest Frasnian. The most critical horizon, the Upper Kellwasser horizon (*linguiformis* Zone), can be clearly traced to distinct layers in both the Coumiac and La Serre sections. Iridium in the La Serre section either is in very low concentrations or is below the detection limits of their study (around 0.01 ng/g). Only three samples show higher iridium concentrations, none of which exceed 110 pg/g (Girard et al. 1993). Neither microtektites nor nickel-rich spinels have been found in the La Serre section. Preliminary data from the Coumiac section, the Frasnian/Famennian GSSP, likewise show no anomalous iridium concentrations or other evidence of impact during the latest Frasnian (Girard et al. 1993).

Given the negative preliminary results of the French team, the Canadian team has been carefully reexamining its samples and analyses for possible sources of contamination in the interim. Goodfellow (1995) feels that contamination can be ruled out and considers the 5,600 pg/g iridium anomaly to be real. He and his colleagues are at present preparing a manuscript to formally report a very large iridium anomaly at the Frasnian/Famennian boundary at La Serre.

These Montagne Noire sections have been determined, by international agreement, to be the best and most complete sections crossing the Frasnian/Famennian boundary in the entire world. The famous and extensively studied Steinbruch Schmidt section in Germany was a very *very* close contender for the Frasnian/Famennian GSSP. Thus one might designate it the second-best section crossing the Frasnian/Famennian boundary on the planet. The biologically critical Upper Kellwasser horizon (*linguiformis* Zone) is clearly recognizable in all. If any evidence is to be found anywhere for impact in the latest Frasnian, it should be present in these sections. Yet, other than the Canadian report (which is still in preparation), no additional geochemical evidence has been found in these sections that gives any indication that an asteroid impact occurred during the latest Frasnian *linguiformis* Zone.

SUMMARY: THE SEARCH CONTINUES

A summary of the results of the worldwide search for evidence of the impact of extraterrestrial bodies on the Earth at or near the Frasnian/Famennian boundary is given in table 9.1. The search is still continu-

Table 9.1. Summary of Physical Evidence Thus Far Obtained Concerning Possible Extraterrestrial Impacts on the Earth During the Late Frasnian and Early Famennian*

Time (years BP):	Impact Evidence or Geochemical Anomaly:
Early *crepida* Zone ca 365.5 Ma	■ (?) 70 km crater at Taihu, China ■ microtektites in China ■ iridium anomaly in Australia ■ negative $\delta^{13}C$ anomaly in Australia
Late *triangularis* Zone ca 366.0 Ma	
Middle *triangularis* Zone ca 366.5 Ma	
Early *triangularis* Zone ca 367.0 Ma FAMENNIAN	■ (?) 52 km crater at Siljan, Sweden ■ microtektites in Belgium
FRASNIAN *linguiformis* Zone ca 367.5 Ma	■ 2 iridium anomalies, at separate horizons, in China ■ possible iridium anomaly in France ■ positive $\delta^{18}O$ anomaly in Germany and China ■ negative $\delta^{13}C$ anomaly in China ■ positive $\delta^{13}C$ anomaly in Germany, France, Austria, Belgium, and Poland ■ positive $\delta^{34}S$ anomaly in Canada and Poland
Late *rhenana* Zone ca 368.0 Ma	■ iridium anomaly in Australia ■ positive $\delta^{13}C$ anomaly in Germany, France, and Austria
Early *rhenana* Zone ca 368.5 Ma	

* Items whose exact zonal position is not currently known are listed with a question mark.

ing at present, even as this book goes to press. Doubtless new information will need to be added to table 9.1 in the very near future. I encourage the reader to keep a pencil handy and to add new notes and jottings to table 9.1 in the months and years ahead. But what can we say at the present time? I consider the microtektite evidence in table 9.1 to be the most conclusive: Two impacts are thus far known to occur in the biological crisis interval. One impact occurs very near the formal Frasnian/Famennian stratigraphic boundary, apparently in the Early *triangularis* Zone. A second impact occurs 1.5 Ma later, in the Early *crepida* Zone.

The microtektites produced by the first impact, in the Early *triangularis* Zone, are thus far known only from Europe, though the search is on for microtektites at this horizon in New York State and elsewhere in the world. An obvious source for the impact melt glass is the 52 km Siljan Crater, located just to the north of the Belgian field sites where the melt glass has been found, and radiometrically dated at or near the Frasnian/Famennian boundary horizon. Eastern North America was also located close to the Siljan impact site in the Late Devonian (see figure 8.5); thus it is likely that microtektites from this impact will eventually be found here as well.

Microtektites from the second impact, in the Early *crepida* Zone, are currently known only from China. Most paleogeographic reconstructions place Australia fairly close to China in the Late Devonian, but no microtektites have been found at the equivalent horizon in the extensively examined Australian sites. It may be that the Early *crepida* impact was smaller than the earlier Early *triangularis* impact and thus did not scatter as much molten material for as great a distance, or that the melt glass has been diagenetically destroyed in the Australian sediments. The Taihu Lake structure is a likely candidate for these tektites, but a precise radiometric date for this structure remains to be determined.

The geochemical evidence remains inconclusive. Iridium anomalies have now been reported from the Late *rhenana, linguiformis,* and Early *crepida* zones. The first and last of these occurs in microbially concentrated horizons. Thus these anomalies are clearly the result, at least in part, of biological activity. The co-occurrence of microtektites with the Early *crepida* anomaly does leave open the possibility that some of the excess iridium at this horizon may have been generated by an impact. In comparison with the Cretaceous/Tertiary iridium anomaly, all the reported Late Devonian anomalies are also very small. All in all, the search for iridium anomalies in the ancient Late Devonian sediments has been frustrating and has yielded no clear signal comparable to that of the Cretaceous/Tertiary boundary.

The stable isotope data are also inconclusive and often conflicting, as in the case of both positive and negative carbon isotope anomalies being reported from different parts of the world in the *linguiformis* Zone (table 9.1). Clearly, something is going on in marine environments during the Frasnian/Famennian crisis interval, but no strong and clear isotopic signal has yet to emerge. The best one can say at present is that sharp and abrupt changes in environmental conditions, producing anomalous jumps in isotopic ratios, occur in the Late *rhenana, linguiformis,* and Early *crepida* zones.

Evidence of impacts at other times in the Late Devonian, and not just within the 3.0 Ma Frasnian/Famennian crisis interval, also exists. The early Frasnian Flynn Creek impact appears to date biostratigraphically very close to the Givetian/Frasnian boundary and may be associated with the ammonoid Frasnes event. The peculiar Spechty Kopf polymictic diamictite sedimentary deposits of the Appalachians may ultimately be related to the late Famennian Charlevoix impact, which may also be associated with the ammonoid Hangenberg event (near the Famennian/Tournaisian boundary). The coincidence in time of these impact events and the Late Devonian stage boundaries is interesting, to say the least (as discussed in chapter 8).

The multiple-impacts hypothesis continues to be the most viable explanation for the ecological and temporal pattern of events that occur during the latest Frasnian and earliest Famennian. A series of impact-generated temperature shocks, rapid drops in temperature on a global scale, is consistent both with the ecological signature of the crisis and with the fact that it is not a solitary event.

From a purely biological perspective, these impacts should have occurred in the Late *rhenana* and *linguiformis* zones. These two zones contain the most intense losses of diversity seen in the Frasnian/Famennian crisis. It is thus puzzling, and not a little frustrating, to note at this point that the best evidence yet produced in the worldwide search for impact, that of the microtektite layers, points to impacts that *both occur after the most critical biological intervals of the Frasnian/Famennian crisis had passed.* Thus these two impacts are not the principal killers in the mass extinction, even though they do occur in the 3.0 Ma span of the Frasnian/Famennian crisis interval.

In July of 1994 the world witnessed the result of the multiple impacts of the fragments of comet Shoemaker-Levy 9 with Jupiter, the largest planet in the solar system. While the images of massive fireballs and ominous black cloud plumes in the Jovian atmosphere are still vivid in our memory it is appropriate to ponder the result of the impact of numerous explosive, volatile-rich objects with the Earth. Objects many times larger than the comet that exploded in the Earth's atmosphere over Tunguska, Siberia, in 1908, flattening trees over an area of 1,000 km^2 with its shock wave but leaving neither a crater nor an iridium anomaly (Rocchia et al. 1990). Objects large enough to reach and explode not in the atmosphere, but in the world's oceans.

Could the solid-body impacts that did occur in the latter half of the Frasnian/Famennian crisis interval (Early *triangularis* and Early *crepida* zones) represent the final phases of a more intense bombardment that began some 0.5 to 1.5 Ma earlier—a bombardment that began with impacts of large, explosive, volatile-rich cometary bodies in the first half of the Frasnian/Famennian crisis interval (Late *rhenana* and *linguiformis* zones)?

The search continues.

X

Extinction, Ecology, and Evolution

E VOLUTION IS a fact. It is an empirical observation, not a theory. Life has evolved in the past and is still evolving today. Visit the laboratory of any population geneticist or ecologist today and you will be shown evolutionary change in the organisms they studied in the modern world. Visit the laboratories or field sites of paleontologists and you will be shown the actual record of the evolution of life in the past.

Imagine a bright red key on the keyboard of a supermegacomputer that, if pressed, would erase the phrase "theory of evolution" from the literature everywhere. I would be the first to press that key. That phrase has caused enormous confusion, especially for the general public. I would replace it with phrases like "theories of why evolution takes place" or "theories of how evolution works" instead, with emphasis on the plural, as that is what is really meant. In fact, the great majority of times you encounter the phrase "the theory of evolution" in the popular literature what is actually meant is "the theory of natural selection."

What is evolution? In Charles Darwin's time, a general definition would have been "descent with modification." And that is not a bad definition. It means that one generation of animals or plants produces the next generation of offspring, which in turn produces the next generation, and so on. This is the descent part. Upon close observation, however, you will notice that each generation is different from the one preceding; that is, the generations are not simply exact copies of each other. This is the modification part. A more modern definition of evolution would be "any change in the gene frequencies of a population with time" (modified from Wilson and Bossert 1971). Both definitions are, however, essentially the same.

Why does evolution occur? This is where theory comes in, and there are many theories of what might cause changes in gene frequencies in populations with time. An older theory, familiar to most people, is the theory of the great French scientist Lamarck, which we might call the "theory of the genetic inheritance of acquired phenotypic characteristics." This is a little long, but it sums up the idea. And it is a very respectable scientific theory of why evolution takes place, it has predictions, and it is subject to test (i.e., it is verifiable). It was tested and found not to work. That is the fate of most scientific theories, but without the constant proposing and testing of theories science would make no progress. Thus Lamarck's theory was, and is, important to science.

When you encounter the headline "Scientists Debate the Theory of Evolution," the immediate implication is that evolution itself is being called into question. This is not the case, and journalists do a disservice to both science and the public in running such headlines. In most instances what is being debated is the applicability of one or more theories or models of how evolution actually takes place, such as the debate concerning the phyletic gradualist model of evolutionary change versus the punctuated equilibrial model. Or the debate between neutralist and selectionist models of evolution.

The most widely subscribed-to theory of how evolution takes place is that of natural selection, first proposed by Charles Darwin. If he had not gotten around to proposing it, Alfred Wallace would have instead; thus it was clearly an idea whose time had come in the 1800s. What is natural selection? A precise, rather pithy definition is the "differential change in genotypic frequencies with time, due to the differential reproductive success of their phenotypes" (modified from Wilson and Bossert 1971). The first part of the definition is simply a restatement of the definition of evolution itself. The real heart of the theory is differential reproductive success. If certain organisms in a population reproduce at a higher rate than other organisms, then the next generation will contain more of their genes than the previous one. And that change in gene frequencies, from generation 1 to generation 2, is by definition evolution. Thus natural selection could clearly drive evolution. The definition of natural selection does not specify what causes differential reproductive success; it simply holds that if it does occur, evolution will result. The next question is obvious: "What could determine the differential reproductive success of differing phenotypes, and why do different animals and plants reproduce at different rates?"

Evolution does not take place in an ecological vacuum. To use a rather gory but simple example (of the "nature blood red in tooth and claw" school so popular in the 1800s), if in their first breeding season one young bull elephant kills its competitor for the attention of the female elephants of the herd, then the reproductive success of the unfortunate bull elephant is obviously zero. Its genes are not passed on to the next generation.

We now realize that natural selection operates in more subtle ways, though immediate death is not ruled out. Organisms must function in their environments, and they must interact with other organisms. The science of ecology is concerned with these functions and interactions. If organisms possess morphologies and behaviors (aspects of their phenotypes) that allow them to function well in their ecological setting, then they are healthy, well fed, and potentially able to devote more time to reproduction. If organisms possess morphologies that do not allow them to function as well—say, they cannot run as fast due to the different structure of their legs, or cannot find their prey or other food as quickly due to the different structure of their eyes and visual system—then they must spend more time simply trying to escape predators and to find food, are generally less healthy, and spend less time and energy in reproduction. The differential functioning of morphologies and phenotypes is the study of adaptation, the subject matter of schools of analysis such as functional and constructional morphology within the disciplines of paleobiology and ecology.

Where does mass extinction fit into the theoretical framework of how evolution works? This question is being hotly debated at the present time, as discussed next.

NATURAL SELECTION OR "ACCIDENTAL SELECTION"?

Biologists have indeed built their advances in evolutionary theory on the Darwinian foundation, not realizing that *the foundation is about to topple* . . .

Hsü 1987:177; italics mine

The "Darwinian foundation" referred to here is the theory of natural selection. Why is the theory of natural selection about to topple? According to Hsü and others, it is toppling because of the impact of mass extinction on our understanding of how evolution works.

The evolutionary change observed during mass extinction crises is currently suggested to be not due to natural selection, at least not as Darwin formulated the theory. The Darwinian "survival of the fittest" during normal periods of environmental change is being argued to become the "survival of the luckiest" during the abnormal environmental shocks associated with mass extinction episodes (Hsü 1987, 1988). A new "theory of mass extinction" is being proposed to replace the theory of natural selection for these catastrophes in Earth history, during which a selective process operates that differentiates not on the basis of differential adaptation ("good genes") but on that of chance ("good luck"; see Gould 1985, Raup 1991).

Specifically, Raup (1991) proposes that evolution could occur in three different modes, which he colorfully calls the "Field of Bullets" mode, the "Fair Game" mode, and the "Wanton Extinction" mode. The "Fair Game" mode he considers to be "selective extinction in a Darwinian sense" in which the best-adapted species differentially survive over time (Raup 1991:188). The opposite of the "Fair Game" is the "Field of Bullets," in which extinction is totally random. Death strikes without any regard to differential adaptation, and all organisms suffer extinction in equal measure. But even here a form of selection appears, as Raup (1991) pointed out. Species with large populations, or higher taxa with great diversity at lower taxonomic levels, should be expected to differentially survive totally random extinction due to sheer numbers alone. Under such a scenario, speciose fossil groups like trilobites and ammonites should weather totally random extinction mechanisms pretty well. Clearly, this is not the case in the Frasnian/Famennian mass extinction. Many speciose higher taxa, indeed some of the most diverse and "successful" taxa (in terms of sheer numbers) of the Frasnian, are those which suffer proportionally higher losses of diversity than other clades (like bivalve molluscs) which are relatively species poor. Thus we can rule out the random "Field of Bullets" mode (or "neutralist" mode, as I prefer to call it, as it is blind to adaptation) of extinction, at least for the Frasnian/Famennian crisis. The pattern of differential survival seen in the extinction event (chapter 6) can only be due to natural selection, right? Maybe not.

Raup (1991) distinguishes a third mode of extinction, which he considers to be in between the extrema of random extinction and natural selection–mediated extinction. This mode he calls "Wanton Extinction," a selective form of extinction in which taxa do differentially survive "but not because they are better adapted to their normal environment" (Raup 1991:188–189). At this point the question of what is "normal" or "abnormal" again arises, a question that was treated in detail previously. Times of mass extinction are clearly not normal (see chapter 2).

What might such a scenario entail, in particular with reference to the Frasnian/Famennian mass extinction? Take, for example, the decimation of tropical ecosystems that occurs during the Frasnian/Famennian crisis. During normal times entire assemblages of species might be well adapted to tropical conditions, both on the land in the sea. Those adaptations could be the result of the sorting effect of millions of years of natural selection, of differential reproduction from generation to generation to generation, of organisms that inhabit the hot low latitudes. Then comes an abnormal catastrophe, a rare event, such as an asteroidal strike. Global temperatures plummet, the previously well-adapted species now experience temperatures below anything experienced by tens of thousands of generations of their ancestors. They perish in droves, and those that do survive do so barely and have little energy to commit to producing the next generation of descendants. Reproduction rates drop, compounding the initial population losses. Genetic lineages terminate, species vanish from the Earth.

In contrast, times might actually improve for high-latitude cold-adapted species. After the initial environmental shock, survivors might experience an expansion of habitat space unknown before and migrate into the lower latitudes that previously excluded them. The catastrophic event is clearly selective: high-latitude species differentially survive, low-latitude species perish. Is this not natural selection? Not in the "normal" sense. Both sets of organisms might be highly adapted to their respective environments; to compare the relative state of adaptation between the two ecological groups is to compare apples with oranges. Only now one group (the low-latitude species) experiences a catastrophic change in their environment, whereas the other (the high-latitude species) experiences a less severe intensification of what it experiences on a day-to-day basis. The selective event is thus a chance event, hence the phrase "the survival of the luckiest."

Chance is chance. In our human experience, a chance event might produce a fortunate result or positive benefit, and most would term such an outcome "good luck." The opposite outcome is often termed "bad luck." I personally dislike the term *luck,* as it is often used by individuals who believe the odds can somehow be made "in their favor," as if there existed some entity ("lady luck") who will bend the rules of probability toward some desired outcome for that individual. To hold such a view is pure superstition and displays a total lack of comprehension of the neutrality of chance. Rather than selection by being in the "right" place at the "right" time, selection by the gambling concept of "luck," I prefer the nongambling concept of selection by accident. We all wish to avoid accidents, yet they occur given even our best precautions, and often when least expected. Rather than playing some cosmic gambling game, the Late Devonian biota are caught up in an unforeseen accident, a very Big Accident—a multicar superhighway chain collision. By accidental selection, low-latitude faunas are decimated in the Frasnian/Famennian crisis. Given another accident, the outcome could have been the reverse. Natural selection gives neither group protection for the accident to come.

My colleague Antoni Hoffman, who sadly is no longer with us to goad me with his wry and cynical humor, would have said that all the preceding is pure nonsense, a totally worthless house of cards. The ultimate reductionist, Hoffman (1989) pointed out that environmental change is environmental change, and whenever the environment changes a biotic response is predicted under the theory of natural selection. Nothing is said about how big or small that change must be. If differential survival is seen to occur, then some type of selection must be at work. As the environmental changes are natural, regardless of how big or small, then that selection is natural selection. "Contrary to Hsü's (1986) assertion then, there is, and can be, no irreconcilable clash between the neo-Darwinian paradigm of evolution and the concept of mass extinctions . . ." (Hoffman 1989:24).

Needless to say, the last word in the matter has yet to be uttered. In an amusing aside (one that raised an appreciative chord from her audience) at an international conference concerning the Frasnian/Famennian mass extinction, the French paleontologist Françoise Bigey quipped: "Bryozoans had both *good genes* and *good luck* . . ." (Bigey 1988:60, italics mine). The debate surrounding extinction by bad genes or bad luck, natural selection or accidental selection, continues among theorists concerned with how evolution actually works.

EXTINCTION AND INNOVATION

The great trick in the history of life apparently lies in being able to survive a mass extinction. If you make it through, *you inherit a world empty of competitors.*

Ward 1992:91; italics mine

The Big Five biotic crises in the history of life on Earth are usually viewed as destructive and as aberrations from the normal process of evolution. The disappearance of large numbers of species can accurately be called destructive, and any selection that chiefly eliminates and seemingly does not allow the evolution of resistance to mass extinction is clearly coarse grained (McGhee 1989a). Even surviving an extinction does not necessarily mean you will do well in the future. For example, the brachiopod spiriferacean species *Mucrospirifer mucronatus* and *Orthospirifer mesastrialis* are extremely abundant and successful in Middle Devonian communities in the Appalachian Sea in eastern North America. They manage to survive both the Taghanic event extinction in the Givetian and the Frasnes event at the Givetian/Frasnian boundary. They inherited a world largely free of competitors in the early Frasnian, only to succumb to extinction in the face of new spiriferacean immigrants who arrive in the Appalachian Sea later in the Frasnian (McGhee 1995). Surviving a major ecological perturbation is thus not an automatic guarantee of future evolutionary success. Equally important is the rate of evolutionary reaction of the surviving species to their new competitor-free world. (See the general overview of this new area of research in Kerr 1994.)

Major ecological perturbations in the course of evolution, however, may play a creative role in the development of life on Earth (Sepkoski 1985; Gould 1985; Jablonski 1986b; Raup 1991). Repeated periods of diversity loss may prevent life from ever reaching a state of equilibrium or saturation and may keep the biosphere in a state of continual flux. Major ecosystem changes, such as the replacement of the nonavian dinosaurs by the mammals, might never have occurred without the magnitude of a perturbation associated with mass extinction. Mass extinctions may repeatedly "reset the clock" of evolution, thus reversing the process of adaptation of species to ever more subtle ecological interactions and increasingly intense biotic pressure. Extinctions may allow the repeated origination of new and less adaptive evolutionary innovations through the course of geologic time. Thus global biotic crises, although decreasing diversity in the short term, may operate to maintain the ecological diversity of life on a geologic timescale. At least that is the theory under the new mass extinction paradigm.

It is difficult to discern any positive evolutionary benefit in the Frasnian/Famennian mass extinction. As predicted, some groups do rapidly radiate in the brave empty new world of the Famennian. New clymeniid ammonoid species rapidly proliferate in Famennian oceans, cyrtospiriferid brachiopod species spread across the sea floors, to name a few. But where are the radically new evolutionary innovations called for earlier? The most novel biological feature, never before seen in the history of Earth, to appear after the crisis is the evolution of seeds in plants. That innovation certainly has had a major impact on subsequent terrestrial ecosystems through the remainder of geologic time. Yet seeds may have evolved anyway; there is at present no proof that the Frasnian/Famennian mass extinction had anything to do with their appearance in the late Famennian.

One thing is clear: The Frasnian/Famennian mass extinction certainly is destructive. The pattern of destruction of evolutionary lineages and ecosystems that occurs in the Frasnian/Famennian crisis has been the subject of this book. The clock of evolution clearly has been reset in the early Famennian. As a matter of fact, it has now been shown to be reset so radically that a few scientists proposed moving the beginning of the Carboniferous back in time, and placing it at the Frasnian/Famennian boundary (which then would become, of course, the Devonian/Carboniferous boundary as well). A "Daylight Savings Time" proposal, on geologic timeframes—"Spring forward" in normal evolution, but "Fall back" in mass extinction.

The most poignant example of this evolutionary "fallback" in progress is the example of the fate of our own earliest land ancestors, the Frasnian amphibians. They had just emerged from the rivers and lakes, making their first trackways along the river banks, clumsily walking on their stubby limbs in the dry air of the new terrestrial world. We have their fossils; we know they were there. And then they are gone, the rocks are empty. No trackways of their stubby feet, no fragments of their bones. Gone.

Only very late in the Famennian do we once again find their fossil bones. The Frasnian/Famennian mass extinction had not totally wiped them out; they had survived after all. And we are here today.

DRAMATIS PERSONAE

The following is a classification of the biota present on the Earth during the Late Devonian mass extinction, as best as is known from the fossil record. Some forms of life became extinct long before the Frasnian/Famennian crisis; others (like ourselves, the mammals) evolved after the crisis had passed into geologic history. Those organisms that witnessed the event, to survive or to perish, are listed here.

Kingdom MONERA: primitive unicellular prokaryotes. May be chemoautotrophic or photoautotrophic; the most ancient forms are anaerobic.

The most important players in the Late Devonian crisis, at least as preserved in the fossil record, are members of the Phylum Cyanobacteria. The simple microbial cyanobacteria, sometimes also referred to as blue-green "algae," are exceedingly hardy forms of life that have changed little for over two billion years. Their organic filaments can bind sediment together to form stromatolites (fossil "agal mats") or oncolites (fossil "algal ovoids"). Stromatolites and oncolites are common in strata following the Frasnian/Famennian extinction; thus the cyanobacteria not only survive but flourish following the crisis.

Kingdom PROTISTA: unicellular and colonial eukaryotes. May be autotrophic or heterotrophic.

   Phylum Chlorophyta (green algae)
   Phylum Rhodophyta (red algae)
   Phylum Dinoflagellata (dinoflagellates)
   Phylum Foraminiferida (foraminiferids)
   Phylum Actinopoda (radiolarians and kin)

The protists suffer heavy diversity losses in the Late Devonian extinction, both within the water column (the phytoplankton) and on the ocean floor (the benthic foraminiferids). The crisis in the phytoplankton is recorded by the massive reduction in "species" of acritarchs, which is an artificial taxonomic group comprised of the resting cysts of algal species. The phyla Haptophyta (calcareous nannoplankton) and Bacillariophyta (diatoms), important elements of the phytoplankton of the world's oceans today, evolved in the Mesozoic and were not present during the Late Devonian crisis.

Kingdom FUNGI: yeast, molds, mushrooms; unicellular to acellular heterotrophs.

Their role in the Late Devonian crisis is unknown, as they have virtually no fossil record.

Kingdom PLANTAE: true plants; multicellular photoautotrophic eukaryotes.

   Phylum Bryophyta (mosses and liverworts)
   Phylum Psilophyta (primitive stem plants)
   Phylum Lycopodophyta (club mosses and kin)
   Phylum Sphenophyta (horsetails and kin)
   Phylum Filicinophyta (ferns)
   Phylum Cycadophyta (primitive seed plants)

The role of the plants in the Frasnian/Famennian extinction is unclear. Major losses in species diversity occur, and significant changes in terrestrial floral species compositions take place, but the exact timing of these events remains murky at present. Some changes appear to start in the earlier Givetian; others seem to take place later in the Famennian. The plant story is exceedingly complex both because of the fragmentary mode of preservation of these organisms and because of the incompleteness of the terrestrial stratigraphic record. The phyla Coniferophyta (conifers), Ginkgophyta (ginkgos), and Angiospermophyta (flowering plants), the first and last of which are important elements of terrestrial floras today, are younger phyla and had not evolved in the Late Devonian.

Kingdom ANIMALIA: true animals; multicellular heterotrophic eukaryotes. The only known animal phylum (with a fossil record) not present during the Frasnian/Famennian crisis is that of the Archaeocyatha. This is because these peculiar cone-shaped organisms had long before died out during the Cambrian Period.

Phylum Porifera:
   Class Demospongea (demosponges)
   Class Calcarea (calcareous sponges)
   Class Hexactinellida (glass sponges)
The sponges are major players in the Late Devonian crisis, both as victims and as survivors. The important reef-building stromatoporoids (believed to be sclerosponges, a type of demosponge) are decimated in the late Frasnian; at the same time the glass sponges survive and flourish in the early Famennian.

Phylum Cnidaria:
   Class Hydrozoa (hydroids)
   Class Scyphozoa (jellyfishes)
   Class Anthozoa (corals and anemones)
Both the colonial and solitary corals are major victims of the Frasnian/Famennian crisis. The fossil record of the other cnidarians is too fragmentary to discern how they fared during the extinction event.

Phylum Ctenophora:
The ctenophores, or "comb jellies," have virtually no fossil record, but we do know they were present in the Frasnian/Famennian crisis as rare fossils of these organisms (which appear similar to cnidarians) are found in the Early Devonian Hunsrückschiefer in Germany.

Phylum Bryozoa:

The colonial bryozoans, or "moss animals," are affected by the Frasnian/Famennian crisis, but not as severely as either the colonial corals or the solitary brachiopods (to which they are related).

Phylum Brachiopoda:
    Class Inarticulata (inarticulates)
    Class Articulata (lamp shells)

The articulate brachiopods, the "shellfish of the Paleozoic," suffer heavy species losses in the Late Devonian. The less diverse and more primitive inarticulates appear to weather the crisis with little change.

Phylum Mollusca:
    Class Monoplacophora (cap shells)
    Class Amphineura (amphineurans)
    Class Scaphopoda (tooth shells)
    Class Cephalopoda (octopi, squids, and kin)
    Class Gastropoda (snails)
    Class Rostroconcha (rostroconchs)
    Class Bivalvia (bivalves: clams, oysters, and kin)
    ?Class Cricoconarida (cone shells)

The molluscs play a mixed role in the Frasnian/Famennian event, somewhat like the sponges. The cephalopods are major victims, while the snails and bivalves appear to suffer little, if at all, during the crisis. The peculiar but numerous cricoconarids—tiny conelike fossils that may not have been molluscs—disappear totally during the Late Devonian. The less diverse classes of monoplacophorans, amphineurans, scaphopods, and rostroconchs are present but have too fragmentary a fossil record to record their response to the crisis.

Phylum Annelida:

The soft-bodied annelid worms rarely leave any fossil remains, but the record of their activity in the form of burrows and trails is abundant in the stratigraphic record. They apparently suffer little diversity loss in the Frasnian/Famennian crisis.

Phylum Onychophora:

The soft bodied onychophorans (which occupy an evolutionary position between annelids and arthropods) are very rarely preserved in the fossil record. Their response to the Late Devonian mass extinction is unknown.

Phylum Arthropoda:
    Subphylum Trilobitomorpha (trilobites and kin)

Subphylum Chelicerata
    Class Merostomata (sea scorpions and kin)
    Class Arachnida (scorpions and spiders)
Subphylum Mandibulata
    Class Crustacea (ostracodes, crabs, and kin)
    Class Myriapoda (centipedes and kin)
    Class Hexapoda (insects)
The arthropods are ecologically diverse in the Late Devonian, being important members of the marine benthos and zooplankton and being present in freshwater and terrestrial land habitats. Most groups suffer diversity losses throughout the Middle and Late Devonian, of which the Frasnian/Famennian crisis appears to have been one of several. Little is known of the fate of the earliest land insects, because of their poor fossil record.

Phylum Echinodermata:
    Subphylum Blastozoa (blastoids and cystoids)
    Subphylum Crinozoa (sea lilies and feather stars)
    Subphylum Echinozoa (sea urchins, cucumbers, and kin)
    Subphylum Asterozoa (starfish and brittle stars)
Also an ecologically diverse group, the echinoderms have various responses to the Frasnian/Famennian crisis. The crinozoans appear to have suffered the most. The last of the cystoids go extinct, a few asterozoan families also, but other groups appear to be little affected. One subphylum, the Homalozoa, became extinct before the Frasnian/Famennian mass extinction.

Phylum Conodonta:
Tiny, eel-shaped swimmers, the conodonts are abundant in Paleozoic oceans. They suffer major diversity losses in the Late Devonian, with only a few species surviving the late Frasnian conodont "crisis horizon."

Phylum Hemichordata:
The hemichordates are minor players in the Late Devonian extinction. Important elements both in the zooplankton and in marine benthos in the earlier Paleozoic (particularly the graptolites), by the Late Devonian they are few in number.

Phylum Chordata:
    Class Agnatha (jawless fishes)
    Class Placodermi (armored fishes)

Class Chondrichthyes (cartilagineous fishes)
Class Acanthodii (spine fin fishes)
Class Osteichthyes (bony fishes)
Class Amphibia (anamniote tetrapods)
The fishes play a major role in the Late Devonian mass extinction, chiefly as victims. The very first amphibians, our oldest land ancestors, were present to witness the event, but their reaction to the crisis is unknown at present because of their sparse fossil record. They first appear in late Frasnian strata, disappear, and reappear in substantial numbers only much later, at the virtual close of the Devonian. Is this coincidence or cause and effect? The advanced amniote tetrapods—classes Reptilia (reptiles), Dinosauria (saurischians, ornithischians, and birds), and Mammalia (mammals)—evolved later and are not present during the Frasnian/Famennian crisis.

The classification of life is an ongoing process; thus classifications vary with time. The one used earlier represents a personally edited synthesis of the extensive, and excellent, treatments that may be found in Clarkson (1993), Kuhn-Schnyder and Rieber (1986), and Stanley (1993).

GLOSSARY

Abyssal: The deep ocean, usually considered as regions at depths of 4,000–6,000 m below the surface.

Aerobic: Oxygen-rich environments, having more than 1.0 mL/L of O_2. May also refer to metabolic processes that involve free oxygen.

Albedo: The reflecting power of a planet; for the Earth, the amount of solar radiation reflected by the atmosphere and surface relative to the total amount of radiation striking the Earth.

Anaerobic: Oxygen-starved environments, having less than 0.1 mL/L of O_2. May also refer to metabolic processes that do not involve free oxygen.

Anoxic: Environments lacking in oxygen.

Autotrophic: The condition of being able to synthesize food material biologically from inorganic sources, using chemical or solar energy.

Bathyal: The mid deep ocean, usually considered as regions at depths of 2,000–4,000 m below the surface.

Benthic: Living on the sea or lake floor, at the bottom of the water column.

Biomass: The amount of living organic matter per a given volume of environmental space.

Bolide: A general term used for an extraterrestrial object that strikes the Earth and explodes, without specifying if the object is a comet, meteor, or asteroid.

BP: Before Present, a datum against which geologic time is measured, where the "present" is taken as the year A.D. 1950.

Brackish: Waters less saline than normal marine, with salinities of from 0.5 to 33 mg/g.

Browsers: For marine organisms, a feeding mode of scraping layers or films of other organisms (bacteria, algae, and so on) off the substratum.

Calcareous: In biology, having mineralized tissues or skeletal elements made of calcium carbonate.

Carnivore: An organism that eats living animals. Often used as a synonym for *predator.*

Chemoautotrophic: The condition of being able to synthesize food using chemical energy rather than solar energy.

Class: A level in the classification of life below the phylum but above the order.

Community: In ecological usage, a multispecies association of organisms that reoccur throughout a given geographic area with essentially constant relative abundances.

Correlation: In geologic usage, determining whether two geographically separated rock formations are of equivalent age or not, using a variety of stratigraphic techniques.

Detritus feeder: An organism that feeds by ingesting sediment and removing any organic material contained as a source of food.

Dysaerobic: Oxygen-poor environments, having from 0.1 to 1.0 mL/L of O_2.

Ecosystem: An ecological unit denoting a biological community and the physical environment that that community inhabits.

Endobenthic: Living within the sediment of the sea or lake floor (cf. *epibenthic, infauna*).

Epibenthic: Living on the surface of the sea or lake floor (cf. *endobenthic*).

Epipelagic: The water column over the open ocean generally corresponding to the neritic realm over the continental shelf.

Eukaryote: Complex cells with a nucleus and organelles.

Eutrophic: Bodies of water with high nutrient concentrations.

Extinction: The termination of a genetic lineage of any taxonomic level. For a species, the disappearance of that gene pool in time due to the death of all its members before they are able to reproduce a descendant generation.

Family: In the classification of life, a level above the genus but below the class.

Food chain: A conceptual description of feeding relationships in nature from autotrophic organisms (primary producers) to herbivores to carnivores (heterotrophic organisms).

Ga: *Giga annum* (plural: *Giga annos*), a billion years (10^9 yr). May refer both to a duration of time and to a date before the present; the latter is formally designated *Ga BP.*

Genus (plural: *genera*): A level in the classification of life above the species but below the family. All organisms are given a two-part scientific name designating the organism's genus and its species.

Grazer: For marine organisms, a feeding mode of consuming large numbers of much smaller organisms, either in the water column or on the substratum.

Guild: In ecology, a group of species that have similar resource requirements.

Halocline: A zone within the water column where the salinity concentration changes sharply and abruptly (cf. *pycnocline, thermocline*).

Herbivore: An organism that eats plants.

Heterotrophic: The condition of being unable to synthesize food biochemically, and thus of being dependent upon the consumption of other organisms for food (cf. *predator, scavenger*).

Hypersaline: Waters having higher salinity than normal marine, with salinities greater than 38 to 40 mg/g.

Infauna: Animals that are endobenthic.

Iridium: An element in the platinum group having the atomic number of 77.

Isotope: Forms of an element with different atomic weights but the same atomic number, due to variation in the number of neutrons.

ka: *kilo annum* (plural: *kilo annos*), one thousand years (10^3 yr). May refer to both a duration of time and to a date before the present; the latter is formally designated *ka BP*.

Kingdom: The highest level in the classification of life. There are five kingdoms of life on Earth (see Appendix).

Larva (plural: *larvae*): A developmental stage in the growth of organisms that undergo metamorphosis.

Lecithotrophic larvae: Larvae that are planktic but that do not feed upon other plankton, as they subsist upon their own yolk from the egg.

Ma: *Mega annum* (plural: *Mega annos*), one million years (10^6 yr). May refer both to a duration of time or a date before the present; the latter is formally designated *Ma BP*.

Marine: Waters having salinities from 33 to 38 mg/g.

Mesic: Environments with moderate amounts of moisture.

Mesopelagic: Oceanic depths below the epipelagic realm, down to about 2,000 m below surface.

mg/g: Milligram per gram; sometimes also referred to as "parts per thousand."

μg/g: Microgram per gram; sometimes also referred to as "parts per million."

Nannoplankton: Very small plankton, less than 20 μm in size.

Nektic: Describes organisms that actively swim.

Neritic: The water column over the continental shelf, from surface to depths of about 200 m in today's oceans.

Niche: A conceptual term that refers to an organism's place and role in the natural world; its feeding type, habitat, reproductive habits, and so on.

ng/g: Nanogram per gram; sometimes also referred to as "parts per billion."

Oligotrophic: Bodies of water with low nutrient concentrations.

Order: A level in the classification of life above the family but below the class.

Pelagic: The water column over the open ocean, seaward of the continental shelf. Depths within the pelagic realm may be indicated by prefix (e.g., abyssopelagic, bathypelagic, and so on).

pg/g: Picogram per gram; sometimes also referred to as "parts per trillion."

Photic zone: The depth within the water column that light penetrates with sufficient concentration to permit photosynthesis.

Photoautotrophic: The condition of being able to synthesize food from simple inorganic compounds by using solar energy; photosynthesis.

Phylum (plural: *phyla*): A level in the classification of life below the kingdom but above the class.

Phytoplankton: Planktic organisms that are photoautotrophic.

Planktic: Describes organisms that passively float.

Planktotrophic larvae: Planktic larvae that feed upon other plankton.

ppm: Parts per million; μg/g concentrations.

ppb: Parts per billion; ng/g concentrations.

ppt: Parts per trillion; pg/g concentrations.

Predator: An organism that eats other living organisms, animal or plant. Sometimes used as a synonym for *heterotroph*, or for *carnivore*, although *carnivory* in the strict sense refers only to the consumption of animals.

Primary producer: An autotroph (cf. *photoautotrophic, chemoautotrophic*).

Prokaryote: Simple cells lacking a nucleus or organelles.

Province: A unit in biogeography; a geographic area with a characteristic assemblage of species.

Pseudoextinction: The termination of a taxon without the termination of a genetic lineage; e.g., a species that evolves into another species.

Pycnocline: A zone within the water column where water density changes sharply and abruptly (cf. *halocline, thermocline*).

Regression: In geologic usage, a lowering of relative sea level and retreat of marine waters from land surfaces.

Scavenger: An organism that eats dead animal or plant tissue.

Sessile: The condition of being immobile. Many sessile marine organisms are directly attached to the substratum.

Siliceous: Having mineralized tissues or skeletal elements made of silica.

Species: (1) A grouping of individual organisms that share the same gene pool (i.e., that can freely interbreed with one another and produce fertile offspring). The species is the fundamental unit of nature. (2) The lowest level in the classification of life. All organisms are given a two-part scientific name consisting of its genus and species designations.

Suspension feeder: An organism that feeds by removing organic matter suspended in the water column.

Taxon (plural: *taxa*): A general term designating all organisms at a given level in the classification of life.

Temperate zone: On the present Earth, the regions between latitudes 23°30' and 66°33' in the Northern and Southern Hemispheres, with variable climates.

Thermocline: A zone within the water column where temperature changes sharply and abruptly (cf. *halocline, pycnocline*).

Transgression: In geologic usage, a rise in relative sea level and advancement of marine waters over land surfaces.

Trophic level: A level within a food chain or trophic pyramid. All organisms in a given trophic level have the same feeding type (e.g., carnivory, herbivory, and so on).

Trophic pyramid: A conceptual description of the relationship between biomass and position within the food chain. In most ecosystems, the biomass of primary producers is greater than that of herbivores, which in turn is greater than that of carnivores.

Tropical zone: On the present Earth, regions below latitudes 23°30' in the Northern and Southern Hemispheres, with hot and humid climates.

Turbidity: The amount of suspended particulate matter (organic or inorganic) per a given volume of seawater.

Vagile: The condition of being mobile, capable of moving from one area to another.

Unconformity: In geologic usage, a gap or hiatus in a stratigraphic sequence such that units of nonsequential age are in contact. These "gaps in the rock record" may be produced by the simple nondeposition of sediments during a time interval, or the later erosion of sediments that were deposited.

Upwelling: The circulation of oceanic bottom waters (usually cold, dense, and nutrient rich) to the surface of the ocean.

Xeric: Dry or desertlike environmental conditions.

Zooplankton: Planktic organisms that are heterotrophic.

REFERENCES

Ahlberg, P. E. 1991. Tetrapod or near-tetrapod fossils from the Upper Devonian of Scotland. *Nature* 354:298–301.

Ahlberg, P. E. and A. R. Milner. 1994. The origin and early diversification of tetrapods. *Nature* 368:507–514.

Alberti, H. 1979. Devonian trilobite biostratigraphy. In M. R. House, C. T. Scrutton, and M. G. Basset, eds., *The Devonian System, Special Papers in Palaeontology* 23:313–324.

Aldridge, R. J. 1988. Extinction and survival in the Conodonta. In G. P. Larwood, ed., *Extinction and Survival in the Fossil Record, Systematics Association Special Volume* 34:231–256.

Algeo, T. J. 1993. The Middle–Late Devonian increase of land plant biomass: Source of a global biogeochemical crisis. *Society of Economic Paleontologists and Mineralogists Abstracts*, p. 38. State College: Pennsylvania State University.

Algeo, T. J. and J. B. Maynard. 1994. Late Devonian black shales and extinction events: "Rooted" in the evolution of vascular land plants? *Geochemical Event Markers in the Phanerozoic, Abstracts; Erlanger Geologische Abhandlungen* 122:3.

Allen, K. C. and D. L. Dineley. 1988. Mid-Devonian to mid-Permian floral and faunal regions and provinces. In A. L. Harris and D. J. Fettes, eds., *The Caledonian–Appalachian Orogen, Geological Society Special Publication* 38:531–548.

Alvarez, L. W., W. Alvarez, F. Asaro, and H. V. Michel. 1980. Extraterrestrial cause for the Cretaceous–Tertiary extinction. *Science* 208:1095–1108.

Alvarez, W., L. W. Alvarez, F. Asaro, and H. V. Michel. 1982. Current status of the impact theory for the terminal Cretaceous extinction. In L. T. Silver and P. H. Schultz, eds., *Geological Implications of Impacts of Large Asteroids and Comets on the Earth; Geological Society of America Special Paper* 190:305–315.

Anderson, T. F. 1990. Temperature from oxygen isotope ratios. In D. E. G. Briggs and P. R. Crowther, eds., *Paleobiology: A Synthesis*, pp. 403–406. Oxford: Blackwell Scientific Publications.

Anderson, T. F. and M. A. Arthur. 1983. Stable isotopes of oxygen and carbon and their application to sedimentologic and paleoenvironmental problems. *Stable Isotopes in Sedimentary Geology, Society of Economic Paleontologists and Mineralogists Short Course Notes* No. 10.

Bakker, R. T. 1977. Tetrapod mass extinctions—a model of the regulation of speciation rates and immigration by cycles of topographic diversity. In A. Hallam, ed., *Patterns of Evolution as Illustrated by the Fossil Record*, pp. 439–468. Amsterdam: Elsevier.

Bambach, R. K. 1985. Classes and adaptive variety: The ecology of diversification in marine faunas through the Phanerozoic. In J. W. Valentine, ed., *Phanerozoic Diversity Patterns: Profiles in Macroevolution*, pp. 191–253. Princeton: Princeton University Press.

Bambach, R. K., C. R. Scotese, and A. M. Ziegler. 1981. Before Pangaea: The geographies of the Paleozoic world. In B. J. Skinner, ed., *Paleontology and Paleoenvironments*, pp. 116–128. Los Altos, CA: William Kaufmann.

Banks, H. 1980. Floral assemblages in the Siluro-Devonian. In D. L. Dilcher and T. N. Taylor, eds., *Biostratigraphy of Fossil Plants*, pp. 1–24. Stroudsberg, Pa.: Dowden, Hutchinson & Ross.

Bayer, U. and G. R. McGhee, Jr. 1986. Cyclic patterns in the Paleozoic and Mesozoic: Implications for time scale calibrations. *Paleoceanography* 1:383–402.

Bayer, U. and G. R. McGhee, Jr. 1989. Periodicity of Devonian sedimentary and biological perturbations: Implications for the Devonian timescale. *Neues Jahrbuch für Geologie und Paläontologie Monatsheft* 1989(1):1–16.

Becker, R. T. 1986. Ammonoid evolution before, during, and after the "Kellwasser Event"—review and preliminary new results. In O. H. Walliser, ed., *Global Bioevents*, Lecture Notes in Earth Sciences, Vol. 8, pp. 181–188. Berlin: Springer-Verlag.

Becker, R. T. 1993. Anoxia, eustatic changes, and Upper Devonian to lowermost Carboniferous global ammonoid diversity. In M. R. House, ed., *The Ammonoidea: Environment, Ecology, and Evolutionary Change; Systematics Association Special Volume* 47:115–163. Oxford: Clarendon Press.

Becker, R. T. and M. R. House. 1994. Kellwasser events and goniatite successions in the Devonian of the Montagne Noire with comments on possible causations. *Courier Forschungsinstitut Senckenberg* 169:45–77.

Becker, R. T., R. Feist, G. Flajs, M. R. House, and G. Klapper. 1989. Frasnian–Famennian extinction events in the Devonian at Coumiac, southern France. *Comptes Rendus de l'Académie des Sciences, Paris* 309(II):259–266.

Becker, R. T., M. R. House, W. T. Kirchgasser, and P. E. Playford. 1991. Sedimentary and faunal changes across the Frasnian/Famennian boundary in the Canning Basin of western Australia. *Historical Biology* 5:183–196.

Becker, R. T., M. R. House, and W. T. Kirchgasser. 1993. Devonian goniatite biostratigraphy and timing of facies movements in the Frasnian of the Canning Basin, western Australia. In E. A. Hailwood and R. B. Kidd, eds., *High Resolution Stratigraphy, Geological Society Special Publication* 70:293–321.

Belka, Z. and J. Wendt. 1992. Conodont biofacies patterns in the Kellwasser Facies (upper Frasnian/lower Famennian) of the eastern Anti-Atlas, Morocco. *Palaeogeography, Palaeoclimatology, Palaeoecology* 91:143–173.

Berner, R. A. 1990. Atmospheric carbon dioxide levels over Phanerozoic time. *Science* 249:1382–1386.

Berry, W. B. H. and P. Wilde. 1978. Progressive ventilation of the oceans—an explanation for the distribution of the Lower Paleozoic black shales. *American Journal of Science* 278:257–275.

Berry, W. B. N., P. Wilde, and M. S. Quinby-Hunt. 1989. Paleozoic (Cambrian through Devonian) anoxitropic biotopes. *Palaeogeography, Palaeoclimatology, Palaeoecology* 74:3–13.

Bhandari, N. 1991. Collisions with Earth over geologic times and their consequences to the terrestrial environment. *Current Science* 61:97–104.

Bigey, F. P. 1986. Biogeography of Devonian Bryozoa. In C. Nielson and G. P. Larwood, eds., *Bryozoa: Ordovician to Recent*, pp. 9–23. Fredensborg, Denmark: Olsen & Olsen.

Bigey, F. P. 1988. Devonian Bryozoa and global events: The Frasnian/Famennian extinction. In N. J. McMillan, A. F. Embry, and D. J. Glass, eds., *Devonian of the World, Canadian Society of Petroleum Geologists Memoir* 14(3):53–62.

Bohor, B. F., E. E. Foord, P. J. Modreski, and D. M. Triplehorn. 1984. Mineralogic evidence for an impact event at the Cretaceous–Tertiary boundary. *Science* 224:867–869.

Bouckaert, J., M. Coen, M. Coen-Aubert, and M. Dusar. 1974. Excursion I. *International Symposium on Belgian Micropaleontological Limits, Namur 1974; Ministry of Economic Affairs, Geological Survey of Belgium.*

Boucot, A. J. 1975. *Evolution and Extinction Rate Controls.* Amsterdam: Elsevier.

Boucot, A. J. 1988. Devonian biogeography: An update. In N. J. McMillan, A. F. Embry, and D. J. Glass, eds., *Devonian of the World, Canadian Society of Petroleum Geologists Memoir* 14(3):211–227.

Boucot, A. J. and J. Gray. 1983. A Paleozoic Pangaea. *Science* 222:571–581.

Boulter, M. C., R. A. Spicer, B. A. Thomas. 1988. Patterns of plant extinction from some palaeobotanical evidence. In G. P. Larwood, ed., *Extinction and Survival in the Fossil Record, Systematics Association Special Volume* 34:1–36.

Boundy-Sanders, S. Q. 1992. Highly calcic, oddly shaped Upper Devonian microtektites from western Yukon Territory. *EOS Transactions, American Geophysical Union Abstracts Supplement* 73(43):328.

Boundy-Sanders, S. Q. 1994. Personal communication.

Brand, U. 1989. Global climatic changes during the Devonian-Mississippian: Stable isotope biogeochemistry of brachiopods. *Palaeogeography, Palaeoclimatology, Palaeoclimatology (Global and Planetary Change Section)* 75:311–329.

Brandt, D. S. and R. J. Elias. 1989. Temporal variations in tempestite thickness may be a geologic record of atmospheric carbon dioxide. *Geology* 17:951–952.

Brasier, M. D. 1988. Foraminiferal extinction and ecological collapse during global biological events. In G. P. Larwood, ed., *Extinction and Survival in the Fossil Record, Systematics Association Special Volume* 34:37–64.

Brenchley, P. J. 1989. The Late Ordovician extinction. In S. K. Donovan, ed., *Mass Extinctions: Processes and Evidence*, pp. 104–132. New York: Columbia University Press.

Bridge, J. S., E. A. Gordon, and R. C. Titus. 1986. Non-marine bivalves and associated burrows in the Catskill magnafacies (Upper Devonian) of New York State. *Palaeogeography, Palaeoclimatology, Palaeoecology* 55:65–77.

Briggs, D. E. G., R. A. Fortey, and E. N. K. Clarkson. 1988. Extinction and fossil record of the arthropods. In G. P. Larwood, ed., *Extinction and Survival in the Fossil Record, Systematics Association Special Volume* 34:171–209.

Buggisch, W. 1972. Zur Geologie und Geochemie der Kellwasserkalke und ihrer begleitenden Sedimente (Unteres Oberdevon). *Abhandlungen des hessischen Landesamt für Bodenforschung* 62:1–68.

Buggisch, W. 1991. The global Frasnian-Famennian "Kellwasser Event." *Geologische Rundschau* 80:49–72.

Caputo, M. V. 1985. Late Devonian glaciation in South America. *Palaeogeography, Palaeoclimatology, Palaeoecology* 51:291–317.

Caputo, M. V. and J. C. Crowell. 1985. Migration of glacial centers across Gondwana during the Paleozoic era. *Geological Society of America Bulletin* 96:1020–1036.

Casier, J-G. 1989. Paléoécologie des ostracodes au niveau de la limite des ètages Frasnien et Famennien á Senzeilles. *Bulletin de l'Institut Royal des Sciences naturelles de Belgique, Sciences de la terre* 59:79–93.

Chadwick, G. H. 1935. Faunal differentiation in the Upper Devonian. *Geological Society of America Bulletin* 46:305–342.

Chaloner, W. G. and A. Sheerin. 1979. Devonian macrofloras. In M. R. House, C. T. Scrutton, and M. G. Bassett, eds., *The Devonian System, Special Papers in Palaeontology* 23:145–161.

Claeys, P. and J.-G. Casier. 1994. Microtektite-like impact glass associated with the Frasnian–Famennian boundary mass extinction. *Earth and Planetary Science Letters* 122:303–315.

Claeys, P., J.-G. Casier and S. V. Margolis. 1992a. A link between microtektites and Late Devonian mass extinctions. *EOS Transactions, American Geophysical Union Abstracts* Supplement 73(43):328.

Claeys, P., J.-G. Casier and S. V. Margolis. 1992b. Microtektites and mass extinctions from the Late Devonian of Belgium: Evidence for a 367 Ma asteroid impact. *Science* 257:1102–1104.

Claeys, P., F. T. Kyte, and J.-G. Casier. 1994. Frasnian–Famennian boundary: Mass extinctions, anoxic oceans, microtektite layers, but not much iridium? In *New Developments Regarding the K/T Event and Other Catastrophes in History, LPI Contribution* 825:22–24. Lunar and Planetary Institute, Houston.

Clarke, A. 1993. Temperature and extinction in the sea: A physiologist's view. *Paleobiology* 19:499–518.

Clarkson, E. N. K. 1993. *Invertebrate Palaeontology and Evolution.* London: Chapman and Hall.

Clube, S. V. M. and W. M. Napier. 1984. The microstructure of terrestrial catastrophism. *Monthly Notices of the Royal Astronomical Society* 211:953–968.

Conaway, C. A., D. J. Over, W. D. Goodfellow, and D. C. Gregoire. 1994. Platinum group element distribution across the Frasnian–Famennian boundary (Upper Devonian), Hanover Shale, western New York State. *Geological Society of America, Abstracts with Program* 26:12.

Copper, P. 1977. Paleolatitudes in the Devonian of Brazil and the Frasnian–Famennian mass extinction. *Palaeogeography, Palaeoclimatology, Palaeoecology* 21:165–207.

Copper, P. 1984. Cold water oceans and the Frasnian–Famennian extinction crisis. *Geological Society of America Abstracts with Programs* 16:10.

Copper, P. 1986. Frasnian–Famennian mass extinction and cold-water oceans. *Geology* 14:835–839.

Copper, P. 1994. Ancient reef ecosystem expansion and collapse. *Coral Reefs* 13:3–11.

Covey, C., S. J. Ghan, J. J. Walton, and P. R. Weissman. 1990. Global environmental effects of impact-generated aerosols: Results from a general circulation model. In V. L. Sharpton and P. D. Ward, eds., *Global Catastrophes in Earth History: An Interdisciplinary Conference on Impacts, Volcanism, and Mass Mortality; Geological Society of America Special Paper* 247:263–270.

Cowie, J. W., W. Ziegler, and J. Remane. 1989. Stratigraphic Commission accelerates progress, 1984 to 1989. *Episodes* 12:79–83.

Crowley, T. J. and G. R. North. 1988. Abrupt climate change and extinction events in Earth history. *Science* 240:996–1002.

Cuffey, R. J. and F. K. McKinney. 1979. Devonian Bryozoa. In M. R. House, C. T. Scrutton, and M. G. Bassett, eds., *The Devonian System, Special Papers in Palaeontology* 23:307–311.

Daeschler, E. B., N. H. Shubin, K. S. Thomson, and W. W. Amaral. 1994. A Devonian tetrapod from North America. *Science* 265:639–642.

Dalziel, I. W. D., L. H. Dalla Salda, and L. M. Gahagan. 1994. Paleozoic Laurentia–Gondwana interaction and the origin of the Appalachian–Andean mountain system. *Geological Society of America Bulletin* 106:243–252.

Dennison, R. 1978. Placodermi. In H. P. Schultze, ed., *Handbook of Paleoichthyology*, Vol. 2. Stuttgart: Gustav Fisher Verlag.

Dennison, R. 1979. Acanthodii. In H. P. Schultze, ed., *Handbook of Paleoichthyology*, Vol. 5. Stuttgart: Gustav Fisher Verlag.

Dietz, R. S. 1961. Astroblemes. *Scientific American* August:50–58.

Dineley, D. L. 1984. *Aspects of a Stratigraphic System: The Devonian.* New York: John Wiley.

Dineley, D. L. 1990. Paleozoic fishing—the Franklinian grounds. In C. R. Harrington, ed., *Canada's Missing Dimension*, pp. 55–80. Ottawa: National Museum of Nature.

Donovan, S. K. 1987. Iridium anomalous no longer? *Nature* 326:331–332.

Downie, C. 1979. Devonian acritarchs. In M. R. House, C. T. Scrutton, and M. G. Bassett, eds., *The Devonian System, Special Papers in Palaeontology* 23:185–188.

Dunn, P. A., K. C. Lohmann, and N. F. Hurley. 1985. C-O isotopic composition of Devono-Carboniferous carbonates of Belgium and Ireland: Evidence of basinal anoxia and global change. *Geological Society of America, Abstracts with Program* 17:569.

Dutro, J. T., Jr. 1981. Devonian brachiopod biostratigraphy. In W. A. Oliver, Jr., and G. Klapper, eds., *Devonian Biostratigraphy of New York,* Vol. 1, pp. 67–82. Washington, D.C.: International Union of Geological Sciences Subcommission on Devonian Stratigraphy.

Dutro, J. T., Jr. 1984. The Frasnian–Famennian extinction event as recorded by Devonian articulate brachiopods in New Mexico. *Geological Society of America Abstracts with Programs* 16:14.

Dyer, B. D., N. N. Lyalikova, D. Murray, M. Doyle, G. M. Kolesov, and W. E. Krumbein. 1989. Role for microorganisms in the formation of iridium anomalies. *Geology* 17:1036–1039.

Erwin, D. H., J. W. Valentine, and J. J. Sepkoski, Jr. 1987. A comparative study of diversification events: The early Paleozoic versus the Mesozoic. *Evolution* 41:1177–1186.

Farsan, N. M. 1986. Frasnian mass extinction—a single catastrophic event or cumulative? In O. H. Walliser, ed., *Global Bio-events,* Lecture Nnotes in Earth Sciences, Vol. 8, pp. 189–197. Berlin: Springer-Verlag.

Faure, G. 1986. *Principles of Isotope Geochemistry.* New York: John Wiley.

Feist, R. and E. Schindler. 1994. Trilobites during the Frasnian Kellwasser Crisis in European Late Devonian cephalopod limestones. *Courier Forschungsinstitut Senckenberg* 169:195–223.

Fischer, A. G. 1984. The two Phanerozoic Supercycles. In W. A. Berggren and J. A. Van Couvering, eds., *Catastrophes and Earth History,* pp. 129–150. Princeton: Princeton University Press.

Fischer, A. G. and Arthur, M. A. 1977. Secular variations in the pelagic realm. *Society of Economic Paleontologists and Mineralogists Special Publication* 25:19–50.

Flessa, K. W. 1990. The "facts" of mass extinction. In V. L. Sharpton and P. E. Ward, eds., *Global Catastrophes in Earth History: An Interdisciplinary Conference on Impacts, Volcanism, and Mass Mortality; Geological Society of America Special Paper* 247:1–7.

Flessa, K. W., H. K. Erben, A. Hallam, K. J. Hsü, H. M. Hüssner, D. Jablonski, D. M. Raup, J. J. Sepkoski, M. E. Soule, W. Sousa, W. Stinnebeck, and G. J. Vermeij. 1986. Causes and consequences of extinction: Group report. In D. M. Raup and D. Jablonski, eds., *Pattern and Process in the History of Life,* pp. 235–257. Berlin: Springer-Verlag.

Fordham, B.G. 1992. Chronometric calibration of mid-Ordovician to Tournasian conodont zones: a compilation from recent graphic-correlation and isotop studies. *Geological Magazine* 129:709–721.

Frakes, L. A. 1979. *Climates Throughout Geologic Time.* Amsterdam: Elsevier.

Frakes, L. A., J. E. Francis, and J. I. Syktus. 1992. *Climate Modes of the Phanerozoic.* Cambridge: Cambridge University Press.

Geitgey, J. E. 1985. Temperature as a factor affecting conodont diversity and distribution. In R. J. Aldridge, R. L. Austin, and M. P. Smith, eds., *Fourth European Conodont Symposium (ECOS IV), Nottingham 1985, Abstracts,* p. 12. Southampton: University of Southampton.

Geldsetzer, H. H. H., W. D. Goodfellow, and D. J. McLaren. 1993. The Frasnian–Famennian extinction event in a stable cratonic shelf setting: Trout River, Northwest Territories, Canada. *Palaeogeography, Palaeoclimatology, Palaeoecology* 104:81–95.

Geldsetzer, H. H. J., W. D. Goodfellow, D. J. McLaren, and M. J. Orchard. 1987. Sulfur-isotope anomaly associated with the Frasnian–Famennian extinction, Medicine Lake, Alberta, Canada. *Geology* 15:393–396.

Geldsetzer, H. H. J., W. D. Goodfellow, D. J. McLaren, and M. J. Orchard. 1988. Reply on "Sulfur-isotope anomaly associated with the Frasnian–Famennian extinction, Medicine Lake, Alberta, Canada." *Geology* 16:87–88.

Gillespie, W. H., G. W. Rothwell, and S. E. Scheckler. 1981. The earliest seeds. *Nature* 293:462–464.

Gilmour, I., W. S. Wolbach, and E. Anders. 1990. Early environmental effects of the terminal Cretaceous impact. In V. L. Sharpton and P. E. Ward, eds., *Global Catastrophes in Earth History: An Interdisciplinary Conference on Impacts, Volcanism, and Mass Mortality; Geological Society of America Special Paper* 247:383–390.

Girard, C., R. Rocchia, R. Feist, L. Froget, and E. Robin. 1993. No evidence of impact at the Frasnian–Famennian boundary in the stratotype area, southern France. *IGCP Projects 293 and 335: Interdisciplinary Conference on Global Boundary Events, Abstracts,* p. 18. International Geological Correlation Program and the Polish Geological Institute, Warsaw, Poland.

Glass, B. P. 1982. Possible correlations between tektite events and climatic change? In L. T. Silver and P. H. Schultz, eds., *Geological Implications of Impacts of Large Asteroids and Comets on the Earth, Geological Society of America Special Paper* 190:251–256.

Glass, B. P. 1990. Tektites and microtektites: Key facts and inferences. *Tectonophysics* 171:393–404.

Gold, T. and S. Soter. 1980. The deep-earth gas hypothesis. *Scientific American* June:154–164.

Gooday, A. J. and G. Becker. 1979. Ostracodes in Devonian biostratigraphy. In M. R. House, C. T. Scrutton, and M. G. Bassett, eds., *The Devonian System, Special Papers in Palaeontology* 23:193–197.

Goodfellow, W. D. 1995. Personal communication.

Goodfellow, W. D., H. H. J. Geldsetzer, D. J. McLaren, M. J. Orchard, and G. Klapper. 1988. The Frasnian–Famennian extinction: Current results and possible causes. In N. J. McMillan, A. F. Embry, and D. J. Glass, eds., *Devonian of the World, Canadian Society of Petroleum Geologists Memoir* 14(3):9–21.

Gordon, W. A. 1975. Distribution by latitude of Phanerozoic evaporite deposits. *Journal of Geology* 83:671–684.

Gould, S. J. 1985. The paradox of the first tier: An agenda for paleobiology. *Paleobiology* 11:2–12.

Gray, J. 1993. Major Paleozoic land plant evolutionary bio-events. *Palaeogeography, Palaeoclimatology, Palaeoecology* 104:153–169.

Grieve, R. A. F. and P. B. Robertson. 1979. The terrestrial cratering record. *Icarus* 38:212–229.

Grieve, R. A. F. and P. B. Robertson. 1987. Terrestrial impact structures. *Geological Survey of Canada Map* 1658A.

Groos-Uffenorde, H. and E. Schindler. 1990. The effect of global events on entomozoacean Ostracoda. In R. Whatley and C. Maybury, eds., *Ostracoda and Global Events*, pp. 101–112. London: Chapman and Hall.

Gutschick, R. C. 1987. The Kentland Dome, Indiana: A structural anomaly. *Geological Society of America Centennial Field Guide, North-Central Section*, pp. 337–342.

Halas, S., A. Balinski, M. Gruszczynski, A. Hoffman, K. Malkowski, and M. Narkiewicz. 1992. Stable isotope record at the Frasnian/Famennian boundary in southern Poland. *Neues Jahrbuch für Geologie und Paläontologie, Monatsheft* 3:129–138.

Hallam, A. 1981. *Facies Interpretation and the Stratigraphic Record*. Oxford: W. H. Freeman.

Hallam, A. 1989. The case for sea-level change as a dominant causal factor in mass extinction of marine invertebrates. *Philosophical Transactions of the Royal Society of London* B325:437–455.

Hallam, A. 1992. *Phanerozoic Sea-Level Changes*. New York: Columbia University Press.

Hallam, A. and A. I. Miller. 1988. Extinction and survival in the Bivalvia. In G. P. Larwood, ed., *Extinction and Survival in the Fossil Record, Systematics Association Special Volume* 34:121–138.

Halstead, L. B. 1988. Extinction and survival of the jawless vertebrates, the Agnatha. In G. P. Larwood, ed., *Extinction and Survival in the Fossil Record, Systematics Association Special Volume* 34:257–267.

Halstead Tarlo, L. B. 1965. Psammosteiformes (Agnatha), a review with descriptions of new material from the Lower Devonian of Poland, I: General Part. *Palaeontologia Polonica* 13:1–135.

Halstead Tarlo, L. B. 1966. Psammosteiformes (Agnatha), a review with descriptions of new material from the Lower Devonian of Poland, II: Systematic Part. *Palaeontologia Polonica* 15:1–168.

Harland, W. B., A. V. Cox, P. G. Llewellyn, C. A. G. Pickton, A. G. Smith and R. Walters. 1982. *A Geologic Time Scale*. Cambridge: Cambridge University Press.

Harland, W. B., R. L. Armstrong, A. V. Cox, L. E. Craig, A. G. Smith, and D. G. Smith. 1989. *A Geologic Time Scale 1989*. Cambridge: Cambridge University Press.

Heckel, P. H. and B. J. Witzke. 1979. Devonian world palaeogeography determined from distribution of carbonates and related lithic palaeoclimatic indicators. In M. R. House, C. T. Scrutton, and M. G. Bassett, eds., *The Devonian System, Special Papers in Palaeontology* 23:99–123.

Herbosch, A., P. Claeys, and F. T. Kyte. 1994. Anoxic event geochemistry at Frasnian–Famennian boundary in Belgium. *Geochemical Event Markers in the Phanerozoic, Abstracts; Erlanger Geologische Abhandlungen* 122:27.

Heymann, D., L. P. Felipe Chibante, R. R. Brooks, W. S. Wolbach, and R. E. Smalley. 1994. Fullerenes in the Cretaceous–Tertiary boundary layer. *Science* 265:645–647.

Hill, D. 1981. Rugosa and Tabulata. In C. Teichert, ed., *Treatise on Invertebrate Paleontology*, Part F, Suppl. 1, pp. 1–762. Boulder, Colorado, and Lawrence, Kansas: Geological Society of America and the University of Kansas Press.

Hladil, J., Z. Kesslevova, and O. Friakova. 1986. The Kellwasser Event in Moravia. In O. H. Walliser, ed., *Global Bio-events*, Lecture Notes in Earth Sciences, Vol. 8, pp. 213–217. Berlin: Springer-Verlag.

Hoffman, A. 1989. Mass extinctions: The view of a sceptic. *Journal of the Geological Society of London* 146:21–35.

Horowitz, A. S. and J. F. Pachut. 1993. Specific, generic, and familial diversity of Devonian bryozoans. *Journal of Paleontology* 67:42–52.

Hörz, F. 1982. Ejecta of the Ries Crater, Germany. In L. T. Silver and P. H. Schultz, eds., *Geological Implications of Impacts of Large Asteroids and Comets on the Earth, Geological Society of America Special Paper* 190:39–55.

Hou, H., H. Zhou, and P. Muchez. 1992. Frasnian–Famennian event stratigraphy in South China. *Fifth International Conference on Global Bioevents (International Geological Correlation Program 216) Abstracts*, p. 52. Universität Göttingen, Göttingen, Germany.

House, M. R. 1967. Fluctuations in the evolution of Palaeozoic invertebrates. In W. B. Harland et al., eds., *The Fossil Record*, pp. 41–54. London: The Geological Society.

House, M. R. 1973. Delimitation of the Frasnian. *Acta Geologica Polonica* 23:1–14.

House, M. R. 1975. Faunas and time in the marine Devonian. *Proceedings of the Yorkshire Geological Society* 40:459–490.

House, M. R. 1979. Biostratigraphy of the early Ammonoidea. In M. R. House, C. T. Scrutton, and M. G. Bassett, eds., *The Devonian System, Special Papers in Palaeontology* 23:263–280.

House, M. R. 1985. Correlation of mid-Palaeozoic ammonoid evolutionary events with global sedimentary perturbations. *Nature* 313:17–22.

House, M. R. 1988a. Extinction and survival in the Cephalopoda. In G. P. Larwood, ed., *Extinction and Survival in the Fossil Record, Systematics Association Special Volume* 34:139–154.

House, M. R. 1988b. International definition of Devonian System boundaries. *Proceedings of the Ussher Society* 7:41–46.

House, M. R. 1989. Ammonoid extinction events. *Philosophical Transactions of the Royal Society of London* B325:307–326.

House, M. R. 1991. Devonian sedimentary microrhythms and a Givetian time scale. *Proceedings of the Ussher Society* 7:392–395.

House, M. R. and W. T. Kirchgasser. 1993. Devonian goniatite biostratigraphy and timing of facies movements in the Frasnian of eastern North America. In E. A. Hailwood and R. B. Kidd, eds., *High Resolution Stratigraphy, Geological Society Special Publication* 70:267–292.

Hsü, K. J. 1987. Replies on "Darwin's three mistakes." *Geology* 15:175–178.

Hsü, K. J. 1988. *The Great Dying*. Orlando: Harcourt Brace Jovanovich.

Hsü, K. J. and J. A. McKenzie. 1990. Carbon-isotope anomalies at era boundaries: Global catastrophes and their ultimate cause. In V. L. Sharpton and P. D. Ward, eds., *Global Catastrophes in Earth History: An Interdisciplinary Conference on Impacts, Volcanism, and Mass Mortality; Geological Society of America Special Paper* 247:61–70.

Hsü, K. J., H. Oberhänsli, J. Y. Gao, S. Shu, C. Hai-hong, and U. Krähenbühl. 1985. "Strangelove ocean" before the Cambrian explosion. *Nature* 316:809–811.

Hubbard, A. E. and N. L. Gilinsky. 1992. Mass extinctions as statistical phenomena: An examination of the evidence using chi-square tests and bootstrapping. *Paleobiology* 18:148–160.

Huddle, J. W. 1963. Conodonts from the Flynn Creek cryptoexplosive structure, Tennessee. *U.S. Geological Survey Professional Paper* 475-C:C55-C57.

Hurley, N. F. and R. Van der Voo. 1987. Paleomagnetism of Upper Devonian reefal limestones, Canning Basin, western Australia. *Geological Society of America Bulletin* 98:138–146.

Hurley, N. F. and R. Van der Voo. 1990. Magnetostratigraphy, Late Devonian iridium anomaly, and impact hypotheses. *Geology* 18:291–294.

Hüssner, V. H. 1983. Die Faunenwende Perm/Trias. *Geologische Rundschau* 72:1–22.

Hut, P., W. Alvarez, W. P. Elder, T. Hansen, E. G. Kauffman, G. Keller, E. M. Shoemaker, and P. R. Weissman. 1987. Comet showers as a cause of mass extinctions. *Nature* 329:118–126.

Izett, G. A. 1991. Tektites in Cretaceous/Tertiary boundary rocks on Haiti and their bearing on the Alvarez impact extinction hypothesis. *Journal of Geophysical Research* 96:20,879–20,905.

Izett, G. A., W. A. Cobban, J. D. Obradovich, and M. J. Kunk. 1993. The Manson impact structure: $^{40}Ar/^{39}Ar$ age and its distal impact ejecta in the Pierre Shale in southeastern South Dakota. *Science* 262:729–732.

Jablonski, D. 1985. Marine regressions and mass extinctions: A test using the modern biota. In J. W. Valentine, ed., *Phanerozoic Diversity Patterns: Profiles in Macroevolution*, pp. 335–354. Princeton: Princeton University Press.

Jablonski, D. 1986a. Causes and consequences of mass extinctions: A comparative approach. In D. K. Elliott, ed., *Dynamics of Extinction*, pp. 183–229. New York: John Wiley.

Jablonski, D. 1986b. Evolutionary consequences of mass extinction. In D. M. Raup and D. Jablonski, eds., *Patterns and Processes in the History of Life*, pp. 313–329. Berlin: Springer-Verlag.

Jablonski, D. 1989. The biology of mass extinction: A palaeontological view. *Philosophical Transactions of the Royal Society of London* B325:357–368.

Jablonski, D. 1991. Extinctions: A paleontological perspective. *Science* 253:754–757.

Jansa, L. F. 1993. Cometary impacts into ocean: Their recognition and the threshold constraint for biological extinctions. *Palaeogeography, Palaeoclimatology, Palaeoecology* 104:271–286.

Jansa, L. F., M.-P. Aubry, and F. M. Gradstein. 1990. Comets and extinctions: Cause and effect? In V. L. Sharpton and P. D. Ward, eds., *Global Catastrophes in Earth History: An Interdisciplinary Conference on Impacts, Volcanism, and Mass Mortality; Geological Society of America Special Paper* 247:223–232.

References 277

Joachimski, M. M. and W. Buggisch. 1992. Carbon isotope shifts at the Frasnian/ Famennian boundary: Evidence for worldwide Kellwasser events? *Fifth International Conference on Global Bioevents (International Geological Correlation Program 216) Abstracts,* pp. 58–59. Universität Göttingen, Göttingen, Germany.

Joachimski, M. M. and W. Buggisch. 1993. Anoxic events in the Late Frasnian—causes of the Frasnian/Famennian faunal crisis? *Geology* 21:675–678.

Joachimski, M. M. and W. Buggisch. 1994. Comparison of inorganic and organic carbon isotope patterns across the Frasnian/Famennian boundary. *Geochemical Event Markers in the Phanerozoic, Abstracts; Erlanger Geologische Abhandlungen* 122:35.

Johnson, J. G. 1974. Extinction of perched faunas. *Geology* 2:479–482.

Johnson, J. G. 1979. Devonian brachiopod biostratigraphy. In M. R. House, C. T. Scrutton, and M. G. Bassett, eds., *The Devonian System, Special Papers in Palaeontology* 23:291–306.

Johnson, J. G. and A. J. Boucot. 1973. Devonian brachiopods. In A. Hallam, ed., *Atlas of Palaeobiogeography,* pp. 89–96. Amsterdam: Elsevier.

Johnson, J. G. and C. A. Sandberg. 1988. Devonian eustatic events in the western United States and their biostratigraphic responses. In N. J. McMillan, A. F. Embry, and D. J. Glass, eds., *Devonian of the World, Canadian Society of Petroleum Geologists Memoir* 14(3):171–178.

Johnson, J. G., G. Klapper, and C. A. Sandberg. 1985. Devonian eustatic fluctuations in Euramerica. *Geological Society of America Bulletin* 96:567–587.

Jones, E. M. and J. W. Kodis. 1982. Atmospheric effects of large body impacts: the first few minutes. In L. T. Silver and P. H. Schultz, eds., *Geological Implications of Impacts of Large Asteroids and Comets on the Earth, Geological Society of America Special Paper* 190:175–186.

Kalvoda, J. 1986. Upper Frasnian and lower Tournaisian events and evolution of calcareous foraminifera—close links to climate changes. In O. H. Walliser, ed., *Global Bio-events, Lecture Notes in Earth Sciences,* Vol. 8, pp. 225–236. Berlin: Springer-Verlag.

Kalvoda, J. 1990. Late Devonian—Early Carboniferous paleobiogeography of benthic foraminifera and climatic oscillations. In E. G. Kauffman and O. H. Walliser, eds., *Extinction Events in Earth History, Lecture Notes in Earth Sciences,* Vol. 30, pp. 183–187. Berlin: Springer-Verlag.

Kalinko, M. K. 1974. Relation between salt content and oil-gas potential of continents and seas. *International Geological Review* 16:759–768.

Katz, D. J. and D. J. Over. 1994. Sedimentology of graded silt laminae associated with the Frasnian–Famennian boundary in the Hanover Shale, western New York State. *Geological Society of America, Abstracts with Program* 26:28.

Kent, D. V., O. Dia, and J. M. A. Sougy. 1984. Paleomagnetism of lower Middle Devonian and upper Proterozoic-Cambrian(?) rock from Mejeria (Mauritania, West Africa). In R. Van der Voo, C. R. Scotese, and N. Bonhommet, eds., *Plate Reconstruction from Paleozoic Paleomagnetism; American Geophysical Union Geodynamics Series* 12:99–115.

Keppie, J. D. 1977. Plate tectonic interpretations of Paleozoic world maps. *Nova Scotia Department of Mines, Paper* 77–3:75.

Kerr, R. A. 1992. Another impact extinction? *Science* 256:1280.

Kerr, R. A. 1993a. Ancient impact-extinction link tightened in Belgium. *Science* 259:175–176.

Kerr, R. A. 1993b. New crater age undercuts killer comets. *Science* 262:659.

Kerr, R. A. 1994. Who profits from ecological disaster? *Science* 266:28–30.

Kirchgasser, W. T., G. C. Baird, and C. E. Brett. 1988. Regional placement of Middle/Upper Devonian (Givetian–Frasnian) boundary in western New York State. In N. J. McMillan, A. F. Embry, and D. J. Glass, eds., *Devonian of the World, Canadian Society of Petroleum Geologists Memoir* 14(3):113–117.

Klapper, G. 1988. The Montagne Noire Frasnian (Upper Devonian) conodont succession. In N. J. McMillan, A. F. Embry, and D. J. Glass, eds., *Devonian of the World, Canadian Society of Petroleum Geologists Memoir* 14(3):449–468.

Klapper, G. and W. Ziegler. 1979. Devonian conodont biostratigraphy. In M. R. House, C. T. Scrutton, and M. G. Bassett, eds., *The Devonian System, Special Papers in Palaeontology* 23:199–224.

Klapper, G., C. A. Sandberg, C. Collinson, J. W. Huddle, R. W. Orr, L. V. Rickard, D. Schumacher, G. Seddon, and T. T. Uyeno. 1971. North American Devonian conodont biostratigraphy. In W. C. Sweet and S. M. Bergstrom, eds., *Symposium on Conodont Biostratigraphy; Geological Society of America Memoir* 127:285–316.

Klapper, G., R. Feist, R. T. Becker, and M. R. House. 1993. Definition of the Frasnian/Famennian Stage boundary. *Episodes* 16:433–441.

Knoll, A. H. 1989. Evolution and extinction in the marine realm: Some constraints imposed by phytoplankton. *Philosophical Transactions of the Royal Society of London* B325:279–290.

Komor, S. C., J. W. Valley, and P. E. Brown. 1988. Fluid-inclusion evidence for impact heating at the Siljan Ring, Sweden. *Geology* 16:711–715.

Kriz, J. 1979. Devonian bivalvia. In M. R. House, C. T. Scrutton, and M. G. Bassett, eds., *The Devonian System, Special Papers in Palaeontology* 23:255–257.

Kuhn-Schnyder, E. and H. Rieber. 1986. *Handbook of Paleozoology.* Baltimore: Johns Hopkins University Press.

Kyte, F. T. 1988. The extraterrestrial component in marine sediments: Description and interpretation. *Paleoceanography* 3:235–247.

Larson, R. L. 1991. Geological consequences of superplumes. *Geology* 19:963–966.

Lebedev, O. A. and J. A. Clack. 1993. Upper Devonian tetrapods from Andreyevka, Tula Region, Russia. *Palaeontology* 36:721–734.

Lethiers, F. 1983. Paléobiogéographie des faunes d'Ostracodes au Dévonien supérieur. *Lethaia* 16:39–49.

Lethiers, F. and R. Feist. 1991. La crise des ostracodes benthiques au passage Frasnien–Famennien de Coumiac (Montagne Noire, France méridionale). *Comptes Rendus de l'Académie des Sciences, Paris* 312(II):1052–1063.

Lewis, J., G. H. Watkins, H. Hartman, and R. Prinn. 1982. Chemical consequences of major impact events on Earth. In L. T. Silver and P. H. Schultz, eds., *Geological Implications of Impacts of Large Asteroids and Comets on the Earth, Geological Society of America Special Paper* 190:215–221.

Linsley, R. M. 1979. Gastropods of the Devonian. In M. R. House, C. T. Scrutton, and M. G. Bassett, eds., *The Devonian System, Special Papers in Palaeontology* 23:249–254.

Long, J. A. 1993. Early-Middle Palaeozoic vertebrate extinction events. In J. A. Long, ed., *Palaeozoic Vertebrate Biostratigraphy and Biogeography*, pp. 54–63. London: Belhaven Press.

Luck, J. M. and K. K. Turekian. 1983. Osmium-187/Osmium-186 in manganese nodules and the Cretaceous–Tertiary boundary. *Science* 222:613–615.

Lütke, F. 1979. Biostratigraphical significance of the Devonian Dacryoconarida. In M. R. House, C. T. Scrutton, and M. G. Bassett, eds., *The Devonian System, Special Papers in Palaeontology* 23:281–289.

McElhinny, M. W. 1973. *Palaeomagnetism and plate tectonics*. Cambridge: Cambridge University Press.

McGhee, G. R., Jr. 1976. Late Devonian benthic marine communities of the central Appalachian Allegheny Front. *Lethaia* 9:111–136.

McGhee, G. R., Jr. 1977. The Frasnian–Famennian (Late Devonian) boundary within the Foreknobs Formation, Maryland and West Virginia. *Geological Society of America Bulletin* 88:806–808.

McGhee, G. R., Jr. 1981a. The Frasnian–Famennian extinctions: A search for extraterrestrial causes. *Bulletin of the Field Museum of Natural History* 52:3–5.

McGhee, G. R., Jr. 1981b. Evolutionary replacement of ecological equivalents in Late Devonian benthic marine communities. *Palaeogeography, Palaeoclimatology, Palaeoecology* 34:267–283.

McGhee, G. R., Jr. 1982. The Frasnian–Famennian extinction event: A preliminary analysis of Appalachian marine ecosystems. In L. T. Silver and P. H. Schultz, eds., *Geological Implications of Impacts of Large Asteroids and Comets on the Earth, Geological Society of America Special Paper* 190:491–500.

McGhee, G. R., Jr. 1984. Tempo of the Frasnian–Famennian biotic crisis. *Geological Society of America, Abstracts with Program* 16:49.

McGhee, G. R., Jr. 1988a. The Late Devonian extinction event: Evidence for abrupt ecosystem collapse. *Paleobiology* 14:250–257.

McGhee, G. R., Jr. 1988b. Evolutionary dynamics of the Frasnian–Famennian extinction event. In N. J. McMillan, A. F. Embry, and D. J. Glass, eds., *Devonian of the World, Canadian Society of Petroleum Geologists Memoir* 14(3):23–28.

McGhee, G. R., Jr. 1989a. Catastrophes in the history of life. In K. C. Allen and D. E. G. Briggs, eds., *Evolution and the Fossil Record*, pp. 26–50. London: Belhaven Press.

McGhee, G. R., Jr. 1989b. The Frasnian–Famennian extinction event. In S. K. Donovan, ed., *Mass Extinctions: Processes and Evidence*, pp. 133–151. New York: Columbia University Press.

McGhee, G. R., Jr. 1990. The Frasnian–Famennian mass extinction record in the eastern United States. In O. H. Walliser and E. G. Kauffman, eds., *Extinction Events in Earth History*, Lecture Notes in Earth Sciences, Vol. 30, pp. 161–168. Berlin: Springer-Verlag.

McGhee, G. R., Jr. 1991. Extinction and diversification in the Devonian Brachiopoda of New York State: No correlation with sea level? *Historical Biology* 5:215–227.

280 References

McGhee, G. R., Jr. 1992. Evolutionary biology of the Devonian Brachiopoda of New York State: No correlation with rate of change of sea level? *Lethaia* 25:165–172.

McGhee, G. R., Jr. 1994. Comets, asteroids, and the Late Devonian mass extinction. *Palaios* 9:513–515.

McGhee, G. R., Jr. 1995. Late Devonian bioevents in the Appalachian Sea: Immigration, extinction, and species replacements. In C. E. Brett and G. C. Baird, eds., *Paleontological Events: Stratigraphic, Ecologic, and Evolutionary Implications.* New York: Columbia University Press, in press.

McGhee, G. R., Jr. 1996. Geometry of evolution in the biconvex Brachiopoda: Morphological effects of mass extinction. *Neues Jahrbuch für Geologie und Paläontologie, Abhandlungen:* in press.

McGhee, G. R., Jr., and U. Bayer. 1985. The local signature of sea-level changes. In U. Bayer and A. Seilacher, eds., *Sedimentary and Evolutionary Cycles,* pp. 98–112. Berlin: Springer-Verlag.

McGhee, G. R., Jr., and J. M. Dennison. 1980. Late Devonian chronostratigraphic correlations between the central Appalachian Allegheny Front and central and western New York. *Southeastern Geology* 21:279–286.

McGhee, G. R., Jr., and R. G. Sutton. 1981. Late Devonian marine ecology and zoogeography of the central Appalachians and New York. *Lethaia* 14:27–43.

McGhee, G. R., Jr., and R. G. Sutton. 1983. Evolution of late Frasnian (Late Devonian) marine environments in New York and the central Appalachians. *Alcheringa* 7:9–21.

McGhee, G. R., Jr., and R. G. Sutton. 1985. Late Devonian marine ecosystems of the lower West Falls Group in New York. In D. L. Woodrow and W. D. Sevon, eds., *The Catskill Delta, Geological Society of America Special Paper* 201:199–209.

McGhee, G. R., Jr., R. Ganapathy, and E. J. Olsen. 1981. Tests of a proposed extraterrestrial cause for the Frasnian–Famennian (Late Devonian) extinction event. *Papers Presented to the Conference on Large Body Impacts and Terrestrial Evolution: Geological, Climatological, and Biological Implications; Lunar and Planetary Institute Contribution* 449:30.

McGhee, G. R., Jr., J. S. Gilmore, C. J. Orth, and E. J. Olsen. 1984. No geochemical evidence for an asteroidal impact at Late Devonian mass extinction horizon. *Nature* 308:629–631.

McGhee, G. R., Jr., C. J. Orth, L. R. Quintana, J. S. Gilmore, and E. J. Olsen. 1986a. The Late Devonian "Kellwasser Event" mass extinction horizon in Germany: No geochemical evidence for a large-body impact. *Geology* 14:776–779.

McGhee, G. R., Jr., C. J. Orth, L. R. Quintana, J. S. Gilmore, and E. J. Olsen. 1986b.Geochemical analyses of the Late Devonian "Kellwasser Event" stratigraphic horizon at Steinbruch Schmidt (F.R.G.). In O. H. Walliser, ed., *Global Bio-events,* Lecture Notes in Earth Sciences, Vol. 8, pp. 219–224. Berlin: Springer-Verlag.

McGhee, G. R., Jr., U. Bayer, and A. Seilacher. 1991. Biological and evolutionary responses to transgressive–regressive cycles. In G. Einsele, W. Ricken, and A. Seilacher, eds., *Cycles and Events in Stratigraphy,* pp. 696–708. Berlin: Springer-Verlag.

McGregor, D. C. 1979. Spores in Devonian stratigraphical correlation. In M. R.

House, C. T. Scrutton, and M. G. Bassett, eds., *The Devonian System, Special Papers in Palaeontology* 23:163–184.

McIntosh, G. C. and D. B. Macurda. 1979. Devonian echinoderm biostratigraphy. In M. R. House, C. T. Scrutton, and M. G. Bassett, eds., *The Devonian System, Special Papers in Palaeontology* 23:331–334.

McKerrow, W. S. 1978. *The Ecology of Fossils.* Cambridge, Mass.: The MIT Press.

McKerrow, W. S. and A. M. Ziegler. 1972. Paleozoic oceans. *Nature* 240:92–94.

McKinney, M. L. 1985. Mass extinction patterns of marine invertebrate groups and some implications for a causal phenomenon. *Paleobiology* 11:227–233.

McKinney, M. L. and C. W. Oyen. 1989. Causation and nonrandomness in biological and geological time series: Temperature as a proximal control of extinction and diversity. *Palaios* 4:3–15.

McLaren, D. J. 1970. Time, life, and boundaries. *Journal of Paleontology* 44:801–815.

McLaren, D. J. 1982. Frasnian–Famennian extinctions. In L. T. Silver and P. H. Schultz, eds., *Geological Implications of Impacts of Large Asteroids and Comets on the Earth, Geological Society of America Special Paper* 190:477–484.

McLaren, D. J. 1983. Bolides and biostratigraphy. *Geological Society of America Bulletin* 94:313–324.

McLaren, D. J. 1984. An Upper Devonian event: Frasnian–Famennian extinctions. *Geological Society of America, Abstracts with Program* 16:49.

McLaren, D. J. 1985. Mass extinction and iridium anomaly in the Upper Devonian of western Australia: A commentary. *Geology* 13:170–172.

McLaren, D. J. and W. D. Goodfellow. 1990. Geological and biological consequences of giant impacts. *Annual Review of Earth and Planetary Sciences* 18:123–171.

Miller, J. D. and D. V. Kent. 1988. Paleomagnetism of the Silurian–Devonian Andreas redbeds: Evidence for an Early Devonian supercontinent? *Geology* 16:195–198.

Mora, C. I., S. G. Driese, and P. G. Seager. 1991. Carbon dioxide in the Paleozoic atmosphere: Evidence from carbon-isotope compositions of pedogenic carbonate. *Geology* 19:1017–1020.

Morel, P. and E. Irving. 1978. Tentative paleocontinental maps for the early Phanerozoic and Proterozoic. *Journal of Geology* 86:535–561.

Morzadec, P. 1992. Evolution des Asteropyginae (Trilobita) et variations eustatiques au Dévonien. *Lethaia* 25:85–96.

Moses, C. O. 1989. A geochemical perspective on the causes and periodicity of mass extinctions. *Ecology* 70:812–823.

Narkiewicz, M. 1992. Sedimentology, conodont biofacies, and stable isotopes at the Frasnian/Famennian boundary in the Janczyce I Section, Holy Cross Mountains, Poland. *Fifth International Conference on Global Bioevents (International Geological Correlation Program 216) Abstracts,* pp. 81–82. Universität Göttingen, Göttingen, Germany.

Narkiewicz, M. and A. Hoffman. 1989. The Frasnian/Famennian transition: The sequence of events in southern Poland and its implications. *Acta Geologica Polonica* 39:13–28.

Newell, N. D. 1967. Revolutions in the history of life. *Geological Society of America Special Paper* 89:63–91.

Newsom, H. E., G. Graup, D. A. Iseri, J. W. Geissman, and K. Keil. 1990. The formation of the Ries Crater, West Germany: Evidence of atmospheric interactions during a larger cratering event. In V. L. Sharpton and P. D. Ward, eds., *Global Catastrophes in Earth History: An Interdisciplinary Conference on Impacts, Volcanism, and Mass Mortality; Geological Society of America Special Paper* 247:195–206.

Nicoll, R. S. and P. E. Playford. 1988. Upper Devonian iridium anomaly and the Frasnian–Famennian boundary in the Canning Basin, western Australia. *Geological Society of Australia, Abstracts* 291:296.

Nicoll, R. S. and P. E. Playford. 1993. Upper Devonian iridium anomalies, conodont zonation and the Frasnian–Famennian boundary in the Canning Basin, western Australia. *Palaeogeography, Palaeoclimatology, Palaeoecology* 104:105–113.

Nitecki, M. H. 1969. Redescription of *Ischadites koenigii* Murchinson 1839. *Fieldiana Geology* 16:341–359.

Novozhilov, N. I. 1961. Dvustvorchatye listonogie Devona (in Russian). *Akademiya Nauk SSSR, Trudy Paleontologicheskii Institut* 81:1–132.

O'Keefe, J. D. and T. J. Ahrens. 1982. The interaction of the Cretaceous/Tertiary bolide with the atmosphere, ocean, and solid Earth. In L. T. Silver and P. H. Schultz, eds., *Geological Implication of Impacts of Large Asteroids and Comets on the Earth, Geological Society of America Special Paper* 190:103–120.

Oliver, W. A., Jr. 1990. Extinctions and migrations of Devonian rugose corals in the Eastern Americas Realm. *Lethaia* 23:167–178.

Oliver, W. A., Jr., and I. Chlupac. 1991. Defining the Devonian: 1979. *Lethaia* 24:119–122.

Oliver, W. A., Jr., and A. E. H. Pedder. 1979. Rugose corals in Devonian stratigraphical correlation. In M. R. House, C. T. Scrutton and M. G. Bassett, eds., *The Devonian System, Special Papers in Palaeontology* 23:233–248.

Oliver, W. A., Jr., and A. E. H. Pedder. 1989. Origins, migrations, and extinctions of Devonian Rugosa on the North American Plate. *Memoirs of the Association of Australasian Palaeontologists* 8:231–237.

Oliver, W. A., Jr., and A. E. H. Pedder. 1994. Crises in the Devonian history of the rugose corals. *Paleobiology* 20:178–190.

Olsen, E. J., G. R. McGhee, J. S. Gilmore, J. D. Knight, and C. J. Orth. 1982. Negative evidence for iridium anomaly at Upper Devonian mass extinction boundary. *Geological Society of America, Abstracts with Program* 14:580.

Öpik, E. J. 1951. Collision probabilities with the planets and the distribution of interplanetary matter. *Proceedings of the Royal Irish Academy, Section A,* 54:165–199.

Ormiston, A. R. and G. Klapper. 1992. Paleoclimate, controls on Upper Devonian source rock sequences, and stacked extinctions. In S. Lidgard and P. R. Crane, eds., *Fifth North American Paleontological Convention Abstracts and Program, Paleontological Society Special Publication* 6:227.

Orth, C. J. 1989. Geochemistry of the bio-event horizons. In S. K. Donovan, ed., *Mass Extinctions: Processes and Evidence*, pp. 37–72. New York: Columbia University Press.

Orth, C. J., M. Attrep, and L. R. Quintana. 1990. Iridium abundance patterns across bio-event horizons in the fossil record. In V. L. Sharpton and P. D. Ward, eds., *Global Catastrophes in Earth History: An Interdisciplinary Conference on Impacts, Volcanism, and Mass Mortality; Geological Society of America Special Paper* 247:45–59.

Over, D. J. 1994. Conodonts and the Frasnian–Famennian boundary (Upper Devonian) within the upper Hanover Shale, northern Appalachian Basin, western New York State. *Geological Society of America, Abstracts with Program* 26:66.

Paul, C. R. C. 1988. Extinction and survival in the echinoderms. In G. P. Larwood, ed., *Extinction and Survival in the Fossil Record, Systematics Association Special Volume* 34:155–170.

Pedder, A. E. H. 1982. The rugose coral record across the Frasnian–Famennian boundary. In L. T. Silver and P. H. Schultz, eds., *Geological Implications of Impacts of Large Asteroids and Comets on the Earth, Geological Society of America Special Paper* 190:485–489.

Playford, P. E., D. J. McLaren, C. J. Orth, J. S. Gilmore, and W. D. Goodfellow. 1984. Iridium anomaly in the Upper Devonian of the Canning Basin, western Australia. *Science* 226:437–439.

Plotnick, R. E. 1983. Patterns in the evolution of the eurypterids. Ph.D. dissertation, University of Chicago.

Pollack, J. B., O. B. Toon, T. P. Ackerman, C. P. McKay, and R. P. Turco. 1983. Environmental effects of an impact-generated dust cloud; implications for the Cretaceous-Tertiary extinctions. *Science* 219:287–290.

Popp, B. N., T. F. Anderson, and P. A. Sandberg. 1986. Brachiopods as indicators of original isotopic compositions in some Paleozoic limestones. *Geological Society of America Bulletin* 97:1262–1269.

Poyarkov, B. V. 1979. *Evolution and Distribution of Devonian Foraminifera.* Moscow: Izdatel'stvo Nauka (in Russian).

Prinn, R. G. and B. Fegley. 1987. Bolide impacts, acid rain, and biospheric traumas at the Cretaceous–Tertiary boundary. *Earth and Planetary Science Letters* 83:1–15.

Quinn, J. F. and P. W. Signor. 1989. Death stars, ecology, and mass extinctions. *Ecology* 70:824–834.

Racki, G. 1990. Frasnian/Famennian event in the Holy Cross Mts, central Poland: Stratigraphic and ecological aspects. In E. G. Kauffman and O. H. Walliser, eds., *Extinction Events in Earth History,* Lecture Notes in Earth Sciences, Vol. 30, pp. 169–181. Berlin: Springer-Verlag.

Raup, D. M. 1979. Size of the Permo-Triassic bottleneck and its evolutionary implications. *Science* 206:217–218.

Raup, D. M. 1982. Biogeographic extinction: A feasibility test. In L. T. Silver and P. H. Schultz, eds., *Geological Implications of Impacts of Large Asteroids and Comets on the Earth; Geological Society of America Special Paper* 190:277–281.

Raup, D. M. 1986. *The Nemesis Affair.* New York: W. W. Norton.

Raup, D. M. 1991a. A kill curve for Phanerozoic marine species. *Paleobiology* 17:37–48.

Raup, D. M. 1991b. *Extinction: Bad Genes or Bad Luck?* New York: W. W. Norton.

Raup, D. M. 1992a. Large-body impact and extinction in the Phanerozoic. *Paleobiology* 18:80–88.

Raup, D. M. 1992b. Large-body impact: The least unlikely cause of pulsed extinction. In S. Lidgard and P. R. Crane, eds., *Fifth North American Paleontological Convention Abstracts and Program, Paleontological Society Special Publication* 6:240.

Raup, D. M. and J. J. Sepkoski, Jr. 1982. Mass extinctions in the marine fossil record. *Science* 215:1501–1503.

Raymond, A. 1992. Vascular land plant diversity in a biogeographic context. In S. Lidgard and P. R. Crane, eds., *Fifth North American Paleontological Convention, Abstracts and Programs; Paleontological Society Special Publication* 6:241.

Raymond, A. 1993. Looking for the Frasnian–Famennian mass extinction on land: Can biogeographic change signal the occurrence and causal mechanisms of mass extinction? *Society of Economic Paleontologists and Mineralogists Abstracts*, p. 43. State College: Pennsylvania State University.

Raymond, A. and C. Metz. 1992. Land plants and the Frasnian-Famennian extinction event. *Geological Society of America, Abstracts with Program* 24:A271.

Raymond, A. and C. Metz. 1995. Laurussian land-plant diversity during the Silurian and Devonian: Mass extinction, sampling bias, or both? *Paleobiology* 21:74–91.

Raymond, A., P. H. Kelley, and C. B. Lutken. 1987. Comment on "Frasnian/Famennian mass extinction and cold-water oceans." *Geology* 15:777.

Rehfeld, U. 1989. Patterns of decline in the Eifelian pelecypod faunas in the Rhenish Slate Mountains (West Germany, Central Europe). *28th International Geological Congress Abstracts* 2:685.

Rhodes, M. C. and C. W. Thayer. 1991. Mass extinctions: Ecological selectivity and primary production. *Geology* 19:877–880.

Richardson, J. B. and S. Ahmed. 1988. Miospores, zonation, and correlation of Upper Devonian sequences from western New York State and Pennsylvania. In N. J. McMillan, A. F. Embry, and D. J. Glass, eds., *Devonian of the World, Canadian Society of Petroleum Geologists Memoir* 14(3):541–558.

Richardson, J. B. and D. C. McGregor. 1986. Silurian and Devonian spore zones of the Old Red Sandstone continent and adjacent regions. *Geological Survey of Canada Bulletin* 364:1–79.

Rickard, L. V. 1975. Correlation of the Silurian and Devonian rocks in New York State. *New York State Museum and Science Service, Map and Chart Series*, No. 24.

Riding, R. 1979. Devonian calcareous algae. In M. R. House, C. T. Scrutton, and M. G. Bassett, eds., *The Devonian System, Special Papers in Palaeontology* 23:141–144.

Rigby, J. K. 1979. Patterns in Devonian sponge distribution. In M. R. House, C. T. Scrutton, and M. G. Bassett, eds., *The Devonian System, Special Papers in Palaeontology* 23:225–228.

Robertson, P. B. 1968. La Malbaie structure, Quebec—a Palaeozoic meteorite impact site. *Meteoritics* 4:89–112.

Robinson, J. M. 1990. Lignin, land plants, and fungi: Biological evolution affecting Phanerozoic oxygen balance. *Geology* 15:607–610.

Rocchia, R., P. Bonté, C. Jéhanno, E. Robin, M. de Angelis, and D. Boclet. 1990. Search for the Tunguska event relics in the Antarctic snow and new estimation of the cosmic iridium accretion rate. In V. L. Sharpton and P. D. Ward, eds., *Global Catastrophes in Earth History; An Interdisciplinary Conference on Impacts, Volcanism, and Mass Mortality: Geological Society of America Special Paper* 247:189–193.

Rolfe, W. D. I. and V. A. Edwards. 1979. Devonian Arthropoda (Trilobita and Ostracoda excluded). In M. R. House, C. T. Scrutton, and M. G. Basset, eds., *The Devonian System, Special Papers in Palaeontology* 23:325–329.

Rondot, J. 1970. La structure de Charlevoix comparée à d'autres impacts météoritiques. *Canadian Journal of Earth Sciences* 7:1194–1202.

Rondot, J. 1971. Impactite of the Charlevoix structure, Quebec, Canada. *Journal of Geophysical Research* 76:5414–5423.

Rondot, J. 1975. L'Astroblème de Charlevoix. *Geos*, Spring:18–20.

Rondot, J. 1979. Reconnaissances géologiques dans Charlevoix-Saguenay. *Ministère des Richesses naturelles du Québec*, DPV-682:1–44.

Rozkowska, M. 1969. Famennian tetracoralloid and heterocoralloid fauna from the Holy Cross Mountains (Poland). *Acta Palaeontologica Polonica* 14:1–187.

Rozkowska, M. 1981. On Upper Devonian habitats of rugose corals. *Acta Palaeontologica Polonica* 25:597–611.

Rudwick, M. J. S. 1976. *The Meaning of Fossils*. New York: Neale Watson.

Rudwick, M. J. S. 1979. The Devonian: A system born from conflict. In M. R. House, C. T. Scrutton, and M. G. Basset, eds., *The Devonian System, Special Papers in Palaeontology* 23:9–21.

Rudwick, M. J. S. 1985. *The Great Devonian Controversey*. Chicago: University of Chicago Press.

Sandberg, C. A. and J. E. Warme. 1993. Conodont dating, biofacies, and catastrophic origin or Late Devonian (early Frasnian) Alamo Breccia, southern Nevada. *Geological Society of America, Abstracts with Program* 25:77.

Sandberg, C. A., W. Ziegler, R. Dreesen, and J. L. Butler. 1988. Part 3: Late Frasnian mass extinction: Conodont event stratigraphy, global changes, and possible causes. *Courier Forschungsinstitut Senckenberg* 102:263–307.

Sartenaer, P. 1968. De l'importance stratigraphique des rhynchonelles Famenniennes situées sous la zone à *Ptychomaletoechia omaliusi* (Gosselet, J. 1877). Sixième note: *Pampoecilorhynchus* n. gen. (1). *Bulletin de l'Institut Royal des Sciences naturelles de Belgique* 44(43):1–36.

Sawlowicz, Z. 1993. Iridium and other platinum-group elements as geochemical markers in sedimentary environments. *Palaeogeography, Palaeoclimatology, Palaeoecology* 104:253–270.

Scheckler, S. E. 1984. Floral changes in the Frasnian and Famennian. *Geological Society of America, Abstracts with Program* 16:62.

Scheckler, S. E. 1986. Geology, floristics and paleoecology of Late Devonian coal swamps from Appalachian Laurentia (U.S.A.). *Annales de la Société géologique de Belgique* 109:209–222.

Schindler, E. 1990a. The late Frasnian (Upper Devonian) Kellwasser crisis. In E. G.

Kauffman and O. H. Walliser, eds., *Extinction Events in Earth History*, Lecture Notes in Earth Sciences, Vol. 30, pp. 150–159. Berlin: Springer-Verlag.

Schindler, E. 1990b. Die Kellwasser-Krise (hohe Frasne-Stufe, Ober-Devon). *Göttinger Arbeiten zur Geologie und Paläontologie* 46:1–115.

Schindler, E. 1993. Event-stratigraphic markers within the Kellwasser Crisis near the Frasnian/Famennian boundary (Upper Devonian) in Germany. *Palaeogeography, Palaeoclimatology, Palaeoecology* 104:115–125.

Schlager, W. 1981. The paradox of drowned reefs and carbonate platforms. *Geological Society of America Bulletin* 92:197–211.

Schopf, T. J. M. 1974. Permo-Triassic extinction: Relation to sea floor spreading. *Journal of Geology* 82:129–143.

Schultz, P. H. and D. E. Gault. 1990. Prolonged global catastrophes from oblique impacts. In V. L. Sharpton and P. D. Ward, eds., *Global Catastrophes in Earth History: An Interdisciplinary Conference on Impacts, Volcanism, and Mass Mortality; Geological Society of America Special Paper* 247:239–261.

Scotese, C. R. 1984. Paleozoic paleomagnetism and the assembly of Pangaea. In R. Van der Voo, C. R. Scotese, and N. Bonhommet, eds., *Plate Reconstruction from Paleozoic Paleomagnetism, American Geophysical Union Geodynamics Series* 12:1–10.

Scotese, C. R. and W. S. McKerrow. 1990. Revised world maps and introduction. In W. S. McKerrow and C. R. Scotese, eds., *Palaeozoic Palaeogeography and Biogeography, Geological Society of London Memoir* 51:1–21.

Scotese, C. R., R. K. Bambach, C. Barton, R. Van der Voo, and A. M. Ziegler. 1979. Paleozoic base maps. *Journal of Geology* 87:217–277.

Scrutton, C. T. 1988. Patterns of extinction and survival in Palaeozoic corals. In G. P. Larwood, ed., *Extinction and Survival in the Fossil Record, Systematics Association Special Volume* 34:65–88.

Sepkoski, J. J., Jr. 1982a. Mass extinctions in the Phanerozoic oceans: A review. In L. T. Silver and P. H. Schultz, eds., *Geological Implications of Impacts of Large Asteroids and Comets on the Earth, Geological Society of America Special Paper* 190:283–289.

Sepkoski, J. J., Jr. 1982b. A compendium of fossil marine families. *Milwaukee Public Museum Contributions to Biology and Geology* 51:1–125.

Sepkoski, J. J., Jr. 1985. Some implications of mass extinction for the evolution of complex life. In M. D. Papagiannis, ed., *The Search for Extraterrestrial Life: Recent Developments*, pp. 223–232. Dordrecht: D. Reidel.

Sepkoski, J. J., Jr. 1986. Phanerozoic overview of mass extinctions. In D. M. Raup and D. Jablonski, eds., *Patterns and Processes in the History of Life*, pp. 277–295. Berlin: Springer-Verlag.

Sepkoski, J. J., Jr. 1992. A compendium of fossil marine animal families, 2nd ed. *Milwaukee Public Museum Contributions in Biology and Geology* 83:1–156.

Sepkoski, J. J., Jr. 1994. What I did with my research career: Or how research on biodiversity yielded data on extinction. In W. Glen, ed., *Mass Extinction Debates: How Science Works in a Crisis*, pp. 132–144. Stanford, Calif.: Stanford University Press.

Shear, W. A., P. M. Bonamo, J. D. Grierson, W. D. I. Rolfe, E. L. Smith, and R. A. Norton. 1984. Early land animals in North America: Evidence from Devonian age arthropods from Gilboa, New York. *Science* 224:492–494.

Sheehan, P. M. and T. A. Hansen. 1986. Detritus feeding as a buffer to extinction at the end of the Cretaceous. *Geology* 14:868–870.

Sheridan, R. E. 1987. Pulsation tectonics as the control of long term stratigraphic cycles. *Paleoceanography* 2:97–118.

Simakov, K. V. 1993. Biochronological aspects of the Devonian–Carboniferous crisis in the regions of the former USSR. *Palaeogeography, Palaeoclimatology, Palaeoecology* 104:127–137.

Simberloff, D. 1974. Permo-Triassic extinctions: Effects of area on biotic equilibrium. *Journal of Geology* 82:267–274.

Smit, J., A. Montanari, N. H. M. Swinburne, W. Alvarez, A. R. Hildebrand, S. V. Margolis, P. Claeys, W. Lowrie, and F. Asaro. 1992. Tektite-bearing, deep-water clastic unit at the Cretaceous/Tertiary boundary in northern Mexico. *Geology* 20:99–103.

Smith, A. G., J. C. Briden, and G. E. Drewry. 1973. Phanerozoic world maps. In N. F. Hughes, ed., *Organisms and Continents Through Time, Special Papers in Palaeontology* 12:1–12.

Sorauf, J. E. and A. E. H. Pedder. 1984. Rugose corals and the Frasnian–Famennian boundary. *Geological Society of America Abstracts with Programs* 16:64.

Sorauf, J. E. and A. E. H. Pedder. 1986. Late Devonian rugose corals and the Frasnian–Famennian crisis. *Canadian Journal of Earth Sciences* 23:1265–1287.

Spinar, Z. V. and Z. Burian. 1972. *Life Before Man.* New York: American Heritage Press.

Stanley, S. M. 1987. *Extinction.* New York: Scientific American Books.

Stanley, S. M. 1993. *Exploring Earth and Life Through Time.* New York: W. H. Freeman and Co.

Stearn, C. W. 1979. Biostratigraphy of Devonian stromatoporoids. In M. R. House, C. T. Scrutton, and M. G. Basset, eds., *The Devonian System, Special Papers in Palaeontology* 23:229–232.

Stearn, C. W. 1987. Effect of the Frasnian–Famennian extinction event on the Stromatoporoids. *Geology* 15:677–679.

Stepanova, G. A., V. G. Khalymbadzha, N. G. Chernysheva, L. G. Petrova, and M. V. Postoyalko. 1985. Boundaries of stages of the Upper Devonian on the South Urals (the eastern slope). *Courier Forschungsinstitut Senckenberg* 75:123–134.

Stock, C. W. 1990. Biogeography of Devonian stromatoporoids. In W. S. McKerrow and C. R. Scotese, eds., *Palaeozoic Palaeogeography and Biogeography, Geological Society Memoir* 12:257–265.

Stock, C. W. 1993. Global versus regional controls on the paleobiogeographic distribution of stromatoporoids in the Devonian of North America. *Society of Economic Paleontologists and Mineralogists Abstracts,* p. 53. State College: Pennsylvania State University.

Streel, M. 1986. Miospore contribution to the upper Famennian–Strunian event stratigraphy. *Annales de la Société géologique de Belgique* 109:75–92.

Streel, M. 1992. Climatic impact on Famennian miospore distribution. *Fifth International Conference on Global Bioevents (International Geological Correlation Program 216) Abstracts*, pp. 108–109. Universität Göttingen, Göttingen, Germany.

Sutton, R. G. and G. R. McGhee, Jr. 1985. The evolution of Frasnian marine 'community-types' of south-central New York. In D. L. Woodrow and W. D. Sevon, eds., *The Catskill Delta, Geological Society of America Special Paper* 201:211–224.

Tappan, H. 1980. *The Paleobiology of Plant Protists*. San Francisco: W. H. Freeman.

Tappan, H. 1981. Extinction or survival and diversification: Patterns of selectivity during Paleozoic crises. *Conference on Large Body Impacts and Terrestrial Evolution: Geological, Climatological, and Biological Implications. Lunar and Planetary Institute Contribution Abstracts* 449:54.

Tappan, H. 1982. Extinction or survival: Selectivity and causes of Phanerozoic crises. In L. T. Silver and P. H. Schultz, eds., *Geological Implications of Impacts of Large Asteroids and Comets on the Earth, Geological Society of America Special Paper* 190:265–276.

Tappan, H. and A. R. Loeblich. 1972. Fluctuating rates of protistan evolution, diversification and extinction. *24th International Geological Congress, Paleontology Section* 7:205–213.

Tappan, H. and A. R. Loeblich. 1973. Evolution of the oceanic plankton. *Earth-Science Reviews* 9:207–240.

Tasch, P. 1963. Evolution of the Branchiopoda. In H. B. Whittington and W. D. I. Rolfe, eds., *Phylogeny and Evolution of Crustacea*, pp. 144–157. Boston: Harvard University.

Taylor, P. D. and G. P. Larwood. 1988. Mass extinctions and the pattern of bryozoan evolution. In G. P. Larwood, ed., *Extinction and Survival in the Fossil Record, Systematics Association Special Volume* 34:99–119.

Teichert, C., B. F. Glenister, and R. E. Crick. 1979. Biostratigraphy of Devonian nautiloid cephalopods. In M. R. House, C. T. Scrutton, and M. G. Bassett, eds., *The Devonian System, Special Papers in Palaeontology* 23:259–262.

Thompson, J. B. and C. R. Newton. 1988. Late Devonian mass extinction: Episodic climatic cooling or warming? In N. J. McMillan, A. F. Embry, and D. J. Glass, eds., *Devonian of the World, Canadian Society of Petroleum Geologists Memoir* 14(3):29–34.

Toomey, D. and B. L. Mamet. 1979. Devonian Protozoa. In M. R. House, C. T. Scrutton,, and M. R. Basset, eds., *The Devonian System, Special Papers in Palaeontology* 23:189–192.

Toon, O. B., J. B. Pollack, T. P. Ackerman, R. P. Turco, C. P. McKay, and M. S. Liu. 1982. Evolution of an impact-generated dust cloud and its effects on the atmosphere. In L. T. Silver and P. H. Schultz, eds., *Geological Implications of Impacts of Large Asteroids and Comets on the Earth, Geological Society of America Special Paper* 190:187–200.

Troitskaya, T. D. 1968. *Devonian Bryozoa of Kazakhstan*. Moscow: Nedra Press (in Russian).

Tsien, H. H. 1980. Les régimes récifaux Dévoniens du Ardenne. *Bulletin de la Société géologique de Belgique* 89:71–102.

Valentine, J. W. 1969. Patterns of taxonomic and ecological structure on the shelf benthos during Phanerozoic time. *Palaeontology* 12:684–709.

Valentine, J. W. and D. Jablonski. 1991. Biotic effects of sea level change: The Pleistocene test. *Journal of Geophysical Research* 96(B4):6873–6878.

Van der Voo, R. 1988. Paleozoic paleogeography of North America, Gondwana, and intervening displaced terrenes: Comparisons of paleomagnetism with paleoclimatology and biogeographical patterns. *Geological Society of America Bulletin* 100:311–324.

Van Valen, L. M. 1984. Catastrophes, expectations, and the evidence. *Paleobiology* 10:121–137.

Veevers, J. J. and C. McA. Powell. 1987. Late Paleozoic glacial episodes in Gondwanaland reflected in transgressive–regressive depositional sequences in Euramerica. *Geological Society of America Bulletin* 98:475–487.

Wadleigh, M. A. and J. Veizer. 1992. 18O/16O and 13C/12C in lower Paleozoic articulate brachiopods: Implications for the isotopic composition of seawater. *Geochimica et Cosmochimica Acta* 56:431–443.

Wallace, M. W., R. R. Keays, and V. A. Gostin. 1991. Stromatolitic iron oxides: Evidence that sea-level changes can cause sedimentary iridium anomalies. *Geology* 19:551–554.

Wang, K. 1992a. Late Devonian (early Famennian) microtektites. *Fifth International Conference on Global Bioevents (International Geological Correlation Program 216) Abstracts*, p. 120. Universität Göttingen, Göttingen, Germany.

Wang, K. 1992b. Glassy microspherules (microtektites) from an Upper Devonian limestone. *Science* 256:1547–1550.

Wang, K. and B. D. E. Chatterton. 1993. Microspherules in Devonian sediments: Origins, geological significance, and contamination problems. *Canadian Journal of Earth Science* 30:1660–1667.

Wang, K. and H. H. J. Geldsetzer. 1992. A Late Devonian impact event (about 1.5 Ma after the Frasnian/Famennian crisis) in south China and western Australia and its association with a probable mass extinction event. *Fifth International Conference on Global Bioevents (International Geological Correlation Program 216) Abstracts*, pp. 118–119. Universität Göttingen, Göttingen, Germany.

Wang, K., M. Attrep, and C. J. Orth. 1993. Global iridium anomaly, mass extinction, and redox change at the Devonian–Carboniferous boundary. *Geology* 21:1071–1074.

Wang, K., H. H. J. Geldsetzer, and B. D. E. Chatterton. 1994. A Late Devonian extraterrestrial impact and extinction in eastern Gondwana: Geochemical, sedimentological, and faunal evidence. In B. O. Dressler, R. A. F. Grieve, and V. L. Sharpton, eds., *Large Meteorite Impacts and Planetary Evolution, Geological Society of America Special Paper* 293:111–120.

Wang, K., C. J. Orth, M. Attrep, B. D. E. Chatterton, H. Hongfei, and H. H. J. Geldsetzer. 1991. Geochemical evidence for a catastrophic biotic event at the Frasnian/Famennian boundary in south China. *Geology* 19:776–779.

Ward, P. D. 1992. *On Methuselah's Trail: Living Fossils and the Great Extinctions.* New York: W. H. Freeman.

Warme, J. E. 1994. Catastrophic Alamo Breccia, Upper Devonian, southeastern Nevada. In *New Developments Regarding the K/T Event and Other Catastrophes in Earth History, LPI Contribution* 825:127–128. Lunar and Planetary Institute, Houston.

Warme, J. E., B. W. Ackman, Yarmanto, and A. K. Chamberlain. 1993. Tha Alamo event: Cataclysmic Devonian breccia, southeastern Nevada. In C. W. Gillespie, ed., *Structural and Stratigraphic Relationships of Devonian Reservoir Rocks, East Central Nevada; Nevada Petroleum Society 1993 Field Conference Guidebook,* Reno, NV, pp. 157–165.

Warren, J. W. and N. A. Wakefield. 1972. Trackways of tetrapod vertebrates from the Upper Devonian of Victoria, Australia. *Nature* 238:469–470.

Webb, G. E. 1994. The Frasnian–Famennian extinction event: Dominance of extrinsic over intrinsic factors in the recovery of reef communities. In *New Developments Regarding the K/T Event and Other Catastrophes in Earth History, LPI Contribution* 825:132–133. Lunar and Planetary Institute, Houston.

Wendt, J. 1988. Condensed carbonate sedimentation in the Late Devonian of the eastern Anti-Atlas (Morocco). *Ecologae geologicae Helveticae* 81:155–173.

Wendt, J. and Z. Belka. 1991. Age and depositional environment of Upper Devonian (early Frasnian to early Famennian) black shales and limestones (Kellwasser Facies) in the eastern Anti-Atlas, Morocco. *Facies* 25:51–90.

Westoll, T. S. 1979. Devonian fish biostratigraphy. In M. R. House, C. T. Scrutton, and M. G. Bassett, eds., *The Devonian System, Special Papers in Palaeontology* 23:341–353.

Wilde, P. and W. B. N. Berry. 1984. Destabilization of the oceanic density structure and its significance to marine "extinction" events. *Palaeogeography, Palaeoclimatology, Palaeoecology* 48:143–162.

Wilde, P. and W. B. N. Berry. 1986. The role of oceanographic factors in the generation of global bio-events. In O. H. Walliser, ed., *Global Bio-events,* Lecture Notes in Earth Sciences, Vol. 8, pp. 75–91. Berlin: Springer-Verlag.

Wilde, P. and W. B. N. Berry. 1988. Comment on "Sulfur-isotope anomaly associated with the Frasnian–Famennian extinction, Medicine Lake, Alberta, Canada." *Geology* 16:86.

Wilson, E. O. and W. H. Bossert. 1971. *A Primer of Population Biology.* Sunderland: Sinauer.

Wolbach, W. S., R. S. Lewis, and E. Anders. 1985. Cretaceous extinctions: Evidence for wildfires and search for meteoritic material. *Science* 230:167–170.

Wolbach, W. S., I. Gilmour, E. Anders, C. J. Orth, and R. R. Brooks. 1988. Global fire at the Cretaceous–Tertiary boundary. *Nature* 334:665–669.

Wolbach, W. S., I. Gilmour, and E. Anders. 1990. Major wildfires at the Cretaceous/Tertiary boundary. In V. L. Sharpton and P. D. Ward, eds., *Global Catastrophes in Earth History: An Interdisciplinary Conference on Impacts, Volcanism, and Mass Mortality, Geological Society of America Special Paper* 247:391–400.

Woodrow, D. L., W. E. Sevon, and J. Richardson. 1990. Evidence of an impact-event in the Appalachian Basin near the Mississippian–Devonian time boundary. *Geological Society of America, Abstracts with Program* 22:A92.

Yan, Z., H. Hou, and L. Ye. 1993. Carbon and oxygen isotope event markers near the Frasnian–Famennian boundary, Luoxiu section, south China. *Palaeogeography, Palaeoclimatology, Palaeoecology* 104:97–104.

Zahnle, K. J. 1990. Atmospheric chemistry by large impacts. In V. L. Sharpton and P. D. Ward, eds., *Global Catastrophes in Earth History: An Interdisciplinary Conference on Impacts, Volcanism, and Mass Mortality, Geological Society of America Special Paper* 247:271–288.

Ziegler, W. 1971. Conodont stratigraphy of the European Devonian. In W. C. Sweet and S. M. Bergstrom, eds., *Symposium on Conodont Biostratigraphy, Geological Society of America Memoir* 127:227–284.

Ziegler, W. 1979. Historical subdivisions of the Devonian. In M. R. House, C. T. Scrutton, and M. G. Basset, eds., *The Devonian System, Special Papers in Palaeontology* 23:23–47.

Ziegler, W. 1984. Conodonts and the Frasnian/Famennian crisis. *Geological Society of America, Abstracts with Program* 16:73.

Ziegler, W. and G. Klapper. 1985. Stages of the Devonian System. *Episodes* 8:104–109.

Ziegler, W. and H. R. Lane. 1987. Cycles in conodont evolution from Devonian to mid- Carboniferous. In R. J. Aldridge, ed., *Palaeobiology of Conodonts*, pp. 147–163. Chichester: Ellis Horwood.

Ziegler, W. and C. A. Sandberg. 1990. The Late Devonian standard conodont zonation. *Courier Forschungsinstitut Senckenberg* 121:1–115.